T0076717

ATLAS OF ORAL MICROBIOLOGY

ATLAS OF ORAL MICROBIOLOGY

FROM HEALTHY MICROFLORA TO DISEASE

Edited by

Xuedong Zhou

Yuqing Li

State Key Laboratory of Oral Diseases, West China School/Hospital of Stomatology, Sichuan University, Chengdu, China

AMSTERDAM • BOSTON • HEIDELBERG • LONDON
NEW YORK • OXFORD • PARIS • SAN DIEGO
SAN FRANCISCO • SINGAPORE • SYDNEY • TOKYO
Academic Press is an imprint of Elsevier

Academic Press is an imprint of Elsevier
32 Jamestown Road, London NW1 7BY, UK
525 B Street, Suite 1800, San Diego, CA 92101-4495, USA
225 Wyman Street, Waltham, MA 02451, USA
The Boulevard, Langford Lane, Kidlington, Oxford OX5 1GB, UK

ISBN: 978-0-12-802234-4

British Library Cataloguing-in-Publication Data
A catalogue record for this book is available from the British Library

Library of Congress Cataloging-in-Publication Data
A catalog record for this book is available from the Library of Congress

For information on all Academic Press publications
visit our website at http://store.elsevier.com/

Typeset by TNQ Books and Journals
www.tnq.co.in

Printed and bound in the United States of America

Contents

Contributors

Lei Cheng
Xingqun Cheng
Meng Deng
Xiaohong Deng
Qin Du
Yang Ge
Qiang Guo
Jinzhi He
Wenxiang Jia
Di Kang
Yu Kuang
Chenglong Li
Mingyun Li
Yan Li
Yuqing Li
Chengcheng Liu
Xianghong Liu
Chuan Lu
Xian Peng
Wei Qiu
Yu Sun
Haohao Wang
Suping Wang
Yan Wang
Yuan Wei
Liying Xiao
Xiaorong Xiao
Xin Xu
Kaiyu Yang
Jumei Zeng
Keke Zhang
Wenling Zhang
Xin Zheng
Han Zhou
Shuangshuang Zhou
Xuedong Zhou
Yuan Zhou
Zhu Zhu

Preface

The human body is home to a diverse range of microorganisms, including a variety of bacteria, archaea, fungi, and viruses, estimated to outnumber human cells in a healthy adult by 10-fold. The importance of characterizing human microbiota for understanding health and disease is highlighted by the recent launch of the Human Microbiome Project (HMP) by the National Institutes of Health. Changes in colonization sites or imbalance of normal microflora may lead to the occurrence of the disease. The mouth harbors a diverse, abundant, and complex microbial community and oral microorganisms are often closely related to many infectious diseases.

This book is keeping up with the advanced edge of the international research field of microbiology. For the first time, it innovatively gives us a complete description of the oral microbial systems according to different oral ecosystems. This book collects a large number of oral microbial pictures, including full-color pictures, colonies photos, and electron microscopy photos. It is by far the most abundant oral microbiology atlas and consists of the largest number of pictures. In the meantime, it also describes in detail a variety of experimental techniques, including microbiological isolation, culture, and identification. It is a monograph with strong practical function.

This book is the first professional and comprehensive color atlas of oral microbiology and it has important academic and application value. The editors and writers have long been engaged in teaching and research work in oral microbiology and oral microecology; they had finished the first color atlas of the oral microbial morphology and ultrastructure, which was written in Chinese. They all have a solid theoretical foundation and rich experience in oral microbiological research.

This exceptional book deserves a broad audience, and it will meet the needs of researchers, clinicians, teachers, and students majoring in biology, stomatology, basic medicine, or clinical medicine. It can also be used to facilitate teaching and international academic exchanges.

Preface

CHAPTER

1

Basic Biology of Oral Microbes

1.1 CYTOLOGICAL BASIS OF MICROORGANISMS

The cell is the fundamental unit of all living organisms. Various subunit structures and chemical substances found on and inside the cell make complex cellular functions possible. Microbes can be divided into three major groups according to their morphological structure, degree of differentiation, and chemical composition: eukaryote, prokaryote, and acellular microorganisms (Figure 1.1(A)–(D)).

1.2 MICROBIAL MORPHOLOGY

Microorganisms, also known as microbes, are tiny organisms that are only visible under an optical microscope or an electron microscope. They are small in size and simple in structure. Microbes reproduce quickly, can tolerate a wide range of environmental conditions, are widely distributed, highly variable, and tend to congregate.

1.2.1 Microbial Size

As many types of microbes exist, they vary widely in size. Generally, the units used to measure microbes are μm and nm. Most cocci are 1 μm in diameter. Bacilli can be further divided into coccobacilli, brevibacteria, and long bacilli, and measure approximately 1–10 μm in length and 0.3–1 μm in width. Spirochetes measure approximately 6–20 μm in length and 0.1–0.2 μm in width. Fungi are several times larger than bacteria. Most viruses are smaller than 150 nm and are only visible under the electron microscope. The same microbes can change in size depending on their environment or age (Figure 1.2).

1.2.2 Microbial Morphology

Different types of microbes have different, but characteristic, shapes. Under suitable conditions, the shape and size of microbes are relatively stable. It is important to know the morphological structure of microbes, as it provides us with a better understanding of microbial physiology, pathogenic mechanisms, antigenic features, and allows us to identify them by species. In addition, knowledge of microbial morphology can be helpful in diagnosing disease and in preventing microbial infections.

1. Bacteria are complex and highly variable microbes. They come in four basic shapes: spherical (cocci), rod-shaped (bacilli), arc-shaped (vibrio), and spiral (spirochete) (Figure 1.3(A)).
2. Fungi are divided into unicellular and multicellular according to the number of cells that make up the organism. Unicellular fungi, such as *Saccharomyces* and other yeast-like fungi, are usually round or oval. Multicellular fungi have hyphae and spores. The hyphae and spores of different fungi are shaped differently (Figure 1.3(B) and (C)).
3. Many viruses are spherical or almost spherical, some are rod-shaped (often seen in plant viruses), filamentous (e.g., freshly isolated influenza virus), bullet-shaped (e.g., rabies virus), brick-shaped (e.g., poxvirus), and tadpole-shaped (e.g., bacteriophage) (Figure 1.3(D)).

1.3 MICROBIAL CELL STRUCTURE

Although different microbes possess different cellular structures, there are certain commonalities within groups of microbes.

1.3.1 Basic Bacterial Structures

The architecture of bacterial cells consists of basic and special structures. Basic structures include the cell wall, cell membrane, cytoplasm, nuclear material, ribosome, plasmid, etc. Special structures, which are only found in some bacteria, include the flagellum, pilus, capsule, spore, etc. (Figure 1.4).

Atlas of Oral Microbiology
http://dx.doi.org/10.1016/B978-0-12-802234-4.00001-X

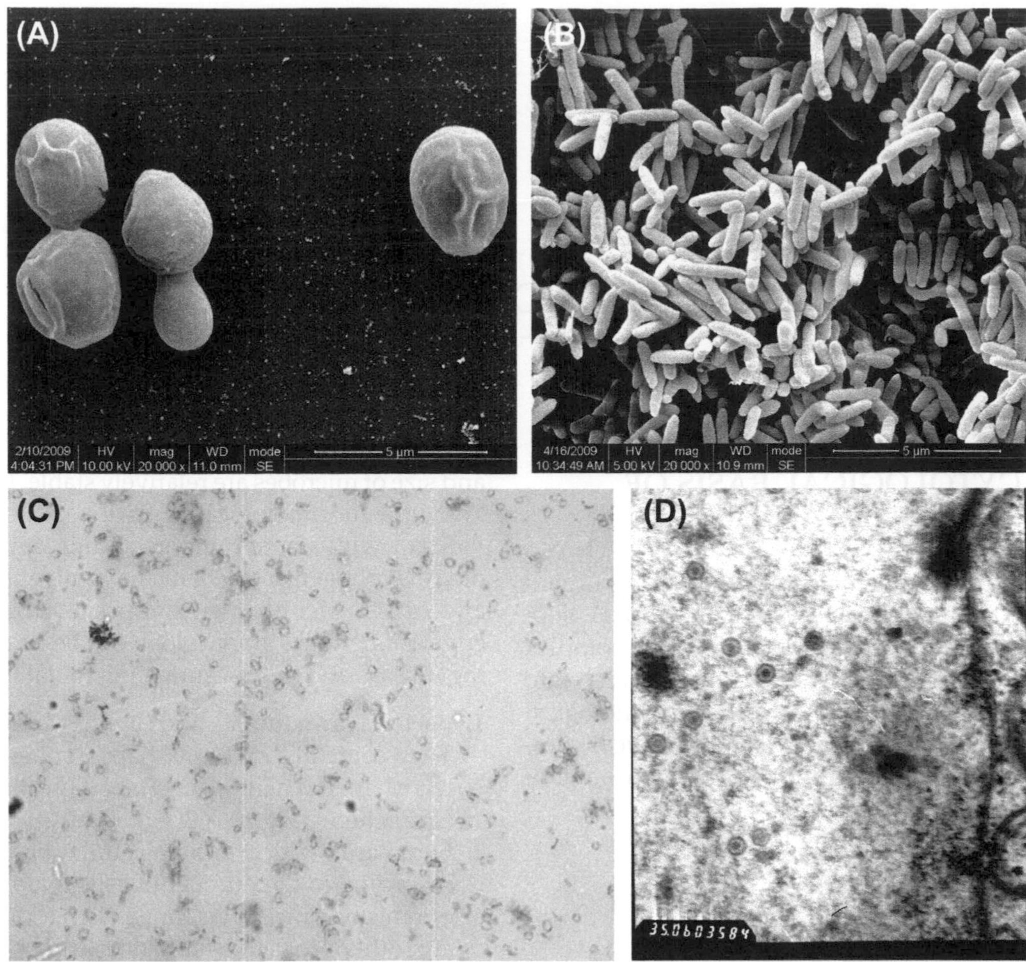

FIGURE 1.1 (A) Eukaryotic microbes (*Saccharomyces*, SEM): The eukaryotic cell has a high degree of nuclear differentiation. It has a nuclear membrane, nucleoli, and chromosomes. There is a complete complement of structured organelles in the cytoplasm and cellular reproduction takes place by mitosis. Examples include fungi and algae. (B) Prokaryotic microbes (bacteria). (C) Prokaryotic microbes (mycoplasma): The prokaryotic cell has a primitive nucleoplasm and cell membrane; it has no nuclear membrane, nucleolus, or organelles. Prokaryotes include bacteria such as *Mycoplasma*, *Chlamydia*, and *Rickettsia*. (D) Acellular microorganisms (herpes simplex virus): Acellular microorganisms are the smallest microorganism with no typical cell structure and no enzymatic energy-production system. They consist merely of a nucleic acid genome (DNA/RNA) and a protein coat (the capsid). Acellular microorganisms can only reproduce inside a living cell. Examples include viruses and subviral agents.

1.3.1.1 Cell Wall

The cell wall is the outermost structure of the bacterial cell and is located outside the cell membrane. It is transparent, tough, and flexible. The average thickness ranges from 15 to 30nm. It mainly consists of peptidoglycan, also called mucopeptide, glycopeptide, or murein. Bacteria are classified into gram-positive and gram-negative based on the appearance of the cells after Gram stain. The peptidoglycan of gram-positive bacteria is composed of a glycan backbone, tetrapeptide side chains, and a pentapeptide cross-linking bridge (Figure 1.5(A)). The peptidoglycan of gram-negative bacteria is composed of a glycan backbone and a tetrapeptide side chain (Figure 1.5(B)).

Gram-positive and gram-negative bacteria have unique structures other than peptidoglycan in their cell walls (Figure 1.5(C) and (D)). Other substances, such as compound polysaccharide, surface protein, proteins M and G of *Streptococcus*, protein A of *Staphylococcus aureus*, etc. are found on the outer layer of the cell wall of some gram-positive bacteria.

1.3.1.2 Cell Wall-Deficient Bacteria (Bacterial L Form)

Cell wall-deficient bacteria are strains of bacteria that lack cell walls. The peptidoglycan that makes up the cell wall can be destroyed or inhibited by physical, chemical, or biological factors. When gram-positive bacteria lack a cell wall, the cytoplasm is surrounded by the cell membrane, and the entire structure is known as a protoplast. When gram-negative bacteria do not have a cell wall, the cytoplasm is protected by the outer membrane, and the entire structure is called a spheroplast. Bacteria

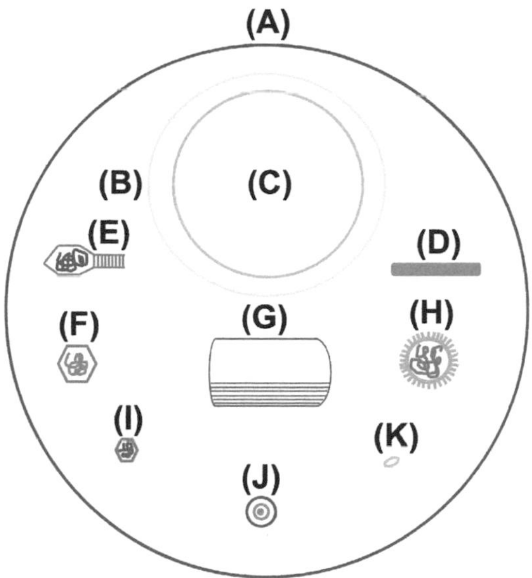

FIGURE 1.2 **Microbial size.** (A) *Staphylococcus* 1000 nm. (B) *Rickettsia* 450 nm. (C) *Chlamydia* 390 nm. (D) Tobacco mosaic virus 300 × 15 nm. (E) *Escherichia coli* bacteriophages, head 65 × 95 nm, tail 12 × 100 nm. (F) Adenovirus 70 nm. (G) Poxvirus 300 × 230 nm. (H) Influenza virus 100 nm. (I) Poliovirus 30 nm. (J) Japanese encephalitis virus 40 nm. (K) Molecule of egg protein 10 nm.

that have lost their cell wall are still capable of growing and dividing as cell wall-deficient bacteria. Examples of these were first isolated in 1935 by Emmy Klieneberger-Nobel, who named them "L-forms" after the Lister Institute in London where she was working at the time. L-form bacteria give rise to a variety of cell morphologies and sizes and can be spherical, rod-shaped, filiform, etc. The rate of growth and division of L-form bacteria is slow. They also form distinctive bacterial colonies when plated on agar. Some L-form strains have a tendency to revert to the normal phenotype when the conditions that were used to produce the cell wall deficiency are reduced. L-form bacteria are difficult to stain or stain unevenly. In a Gram stain test, L-form bacteria always show up as gram-negative, due to the lack of a cell wall.

1.3.1.3 Cell Membrane

The cell membrane is a selectively permeable biological membrane found inside the cell wall and surrounding the cytoplasm. It is made of a lipid bilayer. The cell membrane is compact and flexible, and measures approximately 7.5 nm in thickness. It accounts for 10–30% of the bacterial cell dry weight. The structure of the bacterial cell membrane resembles that of eukaryotic cell membranes, except it is deficient in cholesterol. The lipid bilayer is embedded with carrier proteins and zymoprotein, which possess specific functions.

The cell membrane of some bacteria can form invaginations into the cytoplasm called mesosomes.

1.3.1.4 Cytoplasm

The cytoplasm is the gel-like substance enclosed within the cell membrane, which is made up of water, proteins, lipids, nucleic acids, inorganic salts, etc. Most metabolic activities take place within the cytoplasm, and subcellular structures, such as ribosomes, plasmids, and cytoplasmic granules, are located in the cytoplasm.

Ribosomes are found in cytoplasm. They are approximately 15–20 nm in diameter and are composed of a small (30S) and a large (50S) subunit. The association between subunits requires the presence of Mg^{2+}. Ribosomes are made up of 30% ribosomal proteins and 70% ribosomal RNA.

Plasmids are small, circular, double-stranded DNA molecules and are extrachromosomal genetic material. They can replicate independently of chromosomal DNA and transmit genes encoding drug resistance, bacteriocins, toxins, and more from one bacterium to another via conjugation and transduction.

Cytoplasmic granules is a general term referring to many types of cytoplasmic inclusion granules. They are an intracytoplasmic (inside the cytoplasm of a cell) form of storing nutrients and energy and include molecules such as polysaccharides, lipids, phosphates, etc. They are not essential or permanent structures in cells. Cytoplasmic granules are also known as metachromatic granules because they may stain into different colors than other bacterial cell structures.

1.3.1.5 Nuclear Material

The bacterial nuclear material is also called the nucleoid. It is a piece of double-stranded DNA devoid of nuclear membrane, nucleolus, or histones and is the bacterial equivalent of chromatin. The function of the nucleoid is similar to that of the nucleus in eukaryotic cells and encodes genes necessary for activities and traits such as growth, metabolism, reproduction, heredity, mutation, etc.

1.3.2 Special Bacterial Structures

1.3.2.1 Capsule

The capsule is a layer of slime that lies outside the bacterial cell wall. It is secreted by bacteria and diffuses into the surrounding medium. Based on its appearance when examined by light microscope, the bacterial capsule is classified into two types: microcapsule, which is less than 0.2 μm in thickness and escapes optical detection; and capsule or large capsule, which is over 0.2 μm in thickness, binds tightly to the cell wall, and presents an obvious boundary under optical microscope. The capsule shows up as negatively stained when ordinary

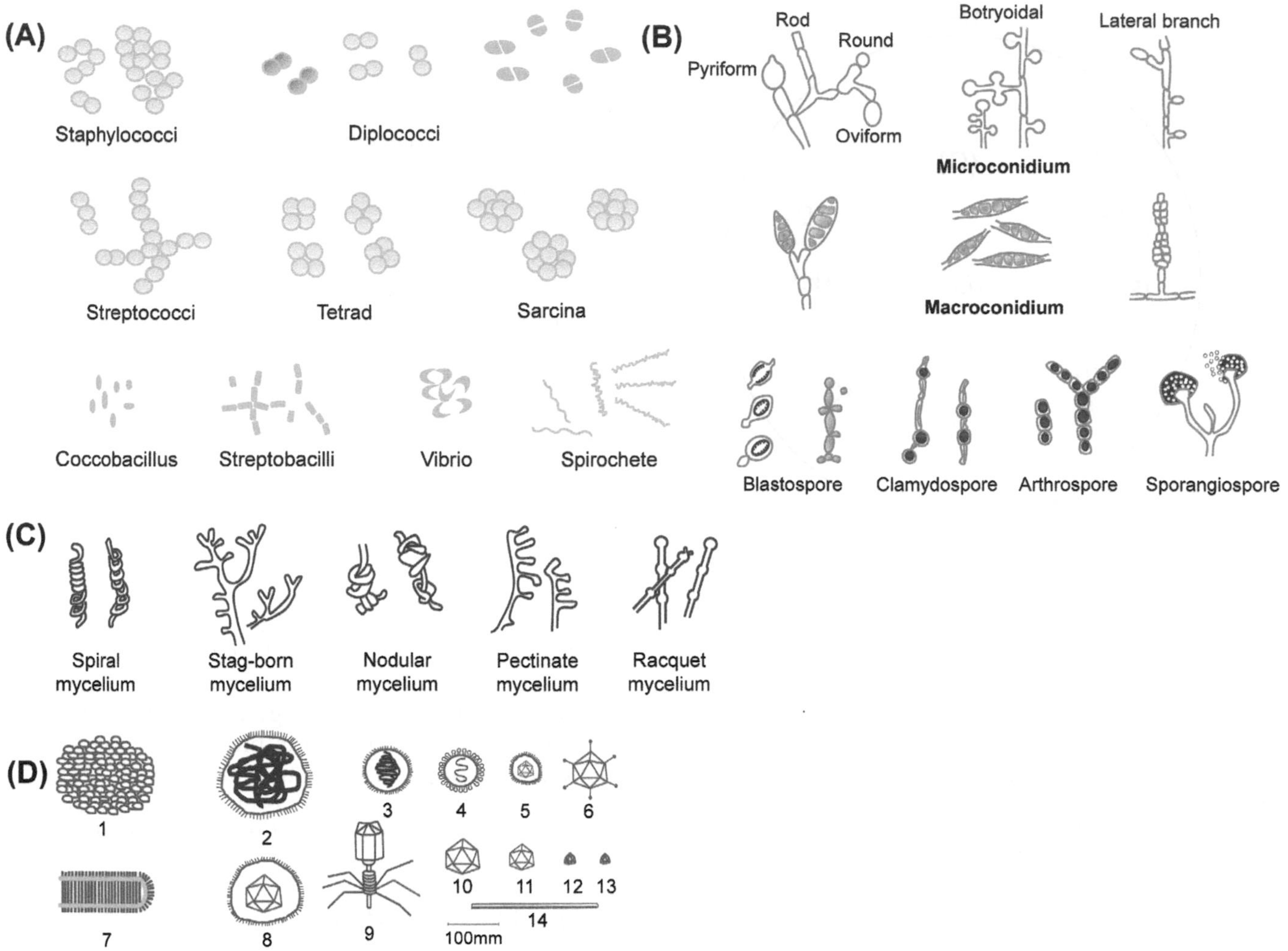

FIGURE 1.3 (A) Basic shape of bacteria. (B) Fungal spores. (C) Fungal hyphae. (D) Morphology and structure of viruses: 1. Poxvirus, 2. Paramyxovirus, 3. Orthomyxovirus, 4. Coronavirus, 5. Togaviridae, 6. Adenovirus, 7. Bullet-shaped virus, 8. Herpes virus, 9. T2 bacteriophage, 10. Reovirus, 11. Papovavirus, 12. Picornavirus, 13. Picodnavirus, 14. Tobacco mosaic virus.

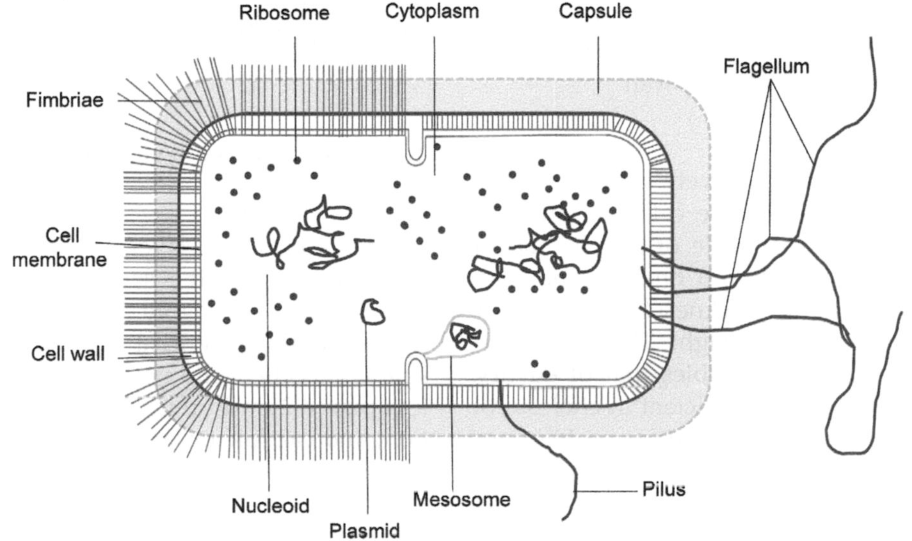

FIGURE 1.4 Schematic representation of bacterial cell structure.

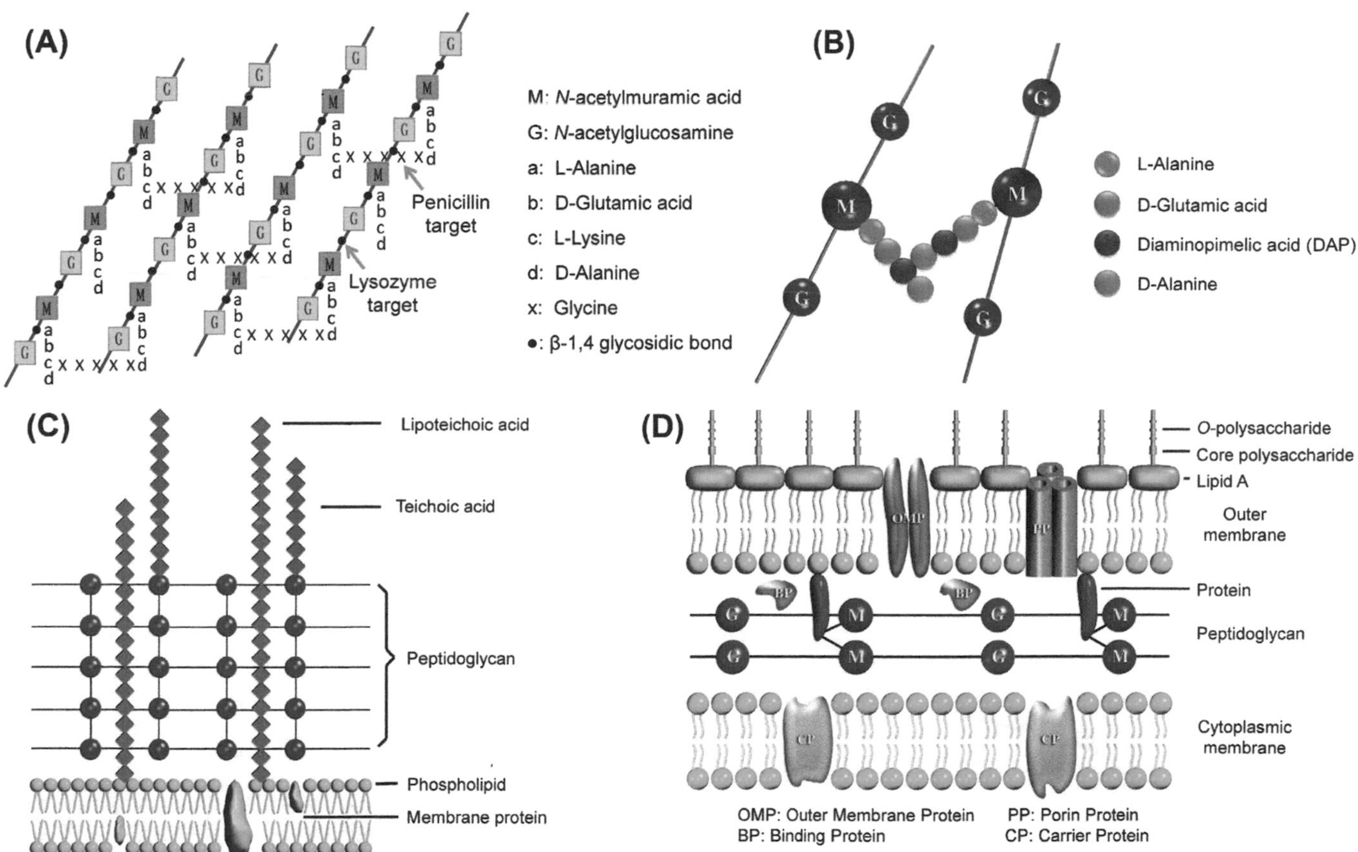

FIGURE 1.5 (A) Schematic representation of *Staphylococcus aureus* cell wall peptidoglycan. M: *N*-Acetylmuramic acid, G: *N*-acetylglucosamine, O: β-1,4 glycosidic bond, a: L-alanine, b: D-aspartic acid, c: L-lysine, d: D-aspartic acid, x: glycine. (B) Schematic representation of *Escherichia coli* cell wall peptidoglycan. M: *N*-Acetylmuramic acid, G: *N*-acetylglucosamine, Ala: alanine, Glu: glutamic acid, DAP: diaminopimelic acid. (C) Schematic of gram-positive bacteria cell wall. Gram-positive bacteria have a thick (20–80 nm) cell wall composed of 15–50 layers of peptidoglycan, many teichoic acids, and some teichuroic acid. Teichoic acids are unique to gram-positive bacterial cell walls and constitute a class of important antigens related to the serotype classification of certain bacterial species. Teichoic acid accounts for about 50% of the dry weight of the cell wall. It is a polymer consisting of ribitol or glycerin residues that are bound by phosphodiester bonds into a long chain, which is then anchored in peptidoglycan. Teichoic acids are classified into cell wall teichoic acids and membrane teichoic acids (also known as lipoteichoic acid, LTA) according to the cellular structure to which they are anchored. (D) Schematic representation of the cell wall of gram-negative bacteria. Gram-negative bacteria have a comparatively thin cell wall, approximately 10–15 nm in thickness. It is made up of one to two layers of peptidoglycan and other complex structures. On the outside of the peptidoglycan is the outer membrane, which is the main component of the gram-negative bacteria cell wall and accounts for approximately 80% of the dry weight of the cell wall. The outer membrane is composed of three layers: lipoprotein, lipid bilayer, and lipopolysaccharide (LPS) ordered from the interior to the exterior. OMP: outer membrane protein, PP: porin, BP: nutrient binding protein, CP: carrier protein, M: *N*-acetylmuramic acid, G: *N*-acetylglucosamine.

staining techniques are used. It appears as a clear halo around the bacterium when stained samples are examined by light microscope. Using special staining, the capsule can be stained differently from the bacterial cell (Figure 1.6). Most bacterial capsules are composed of polysaccharides, but a few capsules are made of polypeptides.

Capsular polysaccharides are highly hydrated molecules in which water accounts for more than 95% of the composition. They bind to phospholipids or lipid A on the cell surface through covalent bonds. The capsule is considered as an important virulence factor because it protects bacteria from engulfment by eukaryotic immune cells, desiccation, and helps bacteria adhere to surfaces.

1.3.2.2 Flagellum

The flagellum is a lash-like appendage that protrudes from the cell body and usually measures 5–20 μm in length and 10–30 nm in diameter. It is the locomotive organelle of motile bacteria such as *Selenomonas* and *Wolinella succinogenes*. The flagellum is composed of three parts: basal body, hook, and filament (Figure 1.7(A)). Different bacteria can have anywhere from one or two flagella to hundreds of flagella (Figure 1.7(B)). Flagella can only be observed directly by electronic microscope or by light microscope after special staining (Figure 1.7(C)). The flagellum is involved in the pathogenesis of some diseases and is antigenic (for example, antigen H). Examples of flagellate bacteria include *Vibrio cholerae*

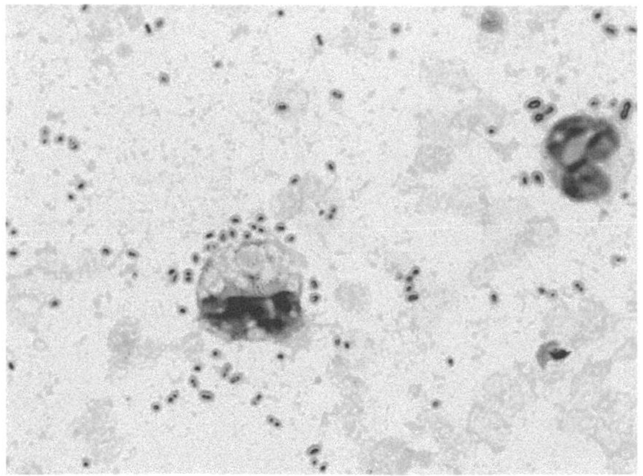

FIGURE 1.6 **Capsule of *S. pneumoniae* (Murs staining method).** Bacterial cells are stained red and the capsule around the cell appears as blue transparent circles.

and *Campylobacter jejuni*, which use multiple flagella to propel themselves through the mucus lining of the small intestine to reach the epithelium and produce toxin.

Flagella can be classified as monotrichous, amphi trichous, lophotrichous, and peritrichous according to their number and location.

1.3.2.3 Pilus

The pilus is a hair-like structure associated with bacterial adhesion and related to bacterial colonization and infection. Pili are primarily composed of oligomeric pilin proteins, which arrange helically to form a cylinder. New pilin protein molecules insert into the base of the pilus. Pili are antigenic, and genes encoding pili can be located in the bacterial chromosome or in plasmids. Pili are not locomotive structures. They are classified into ordinary

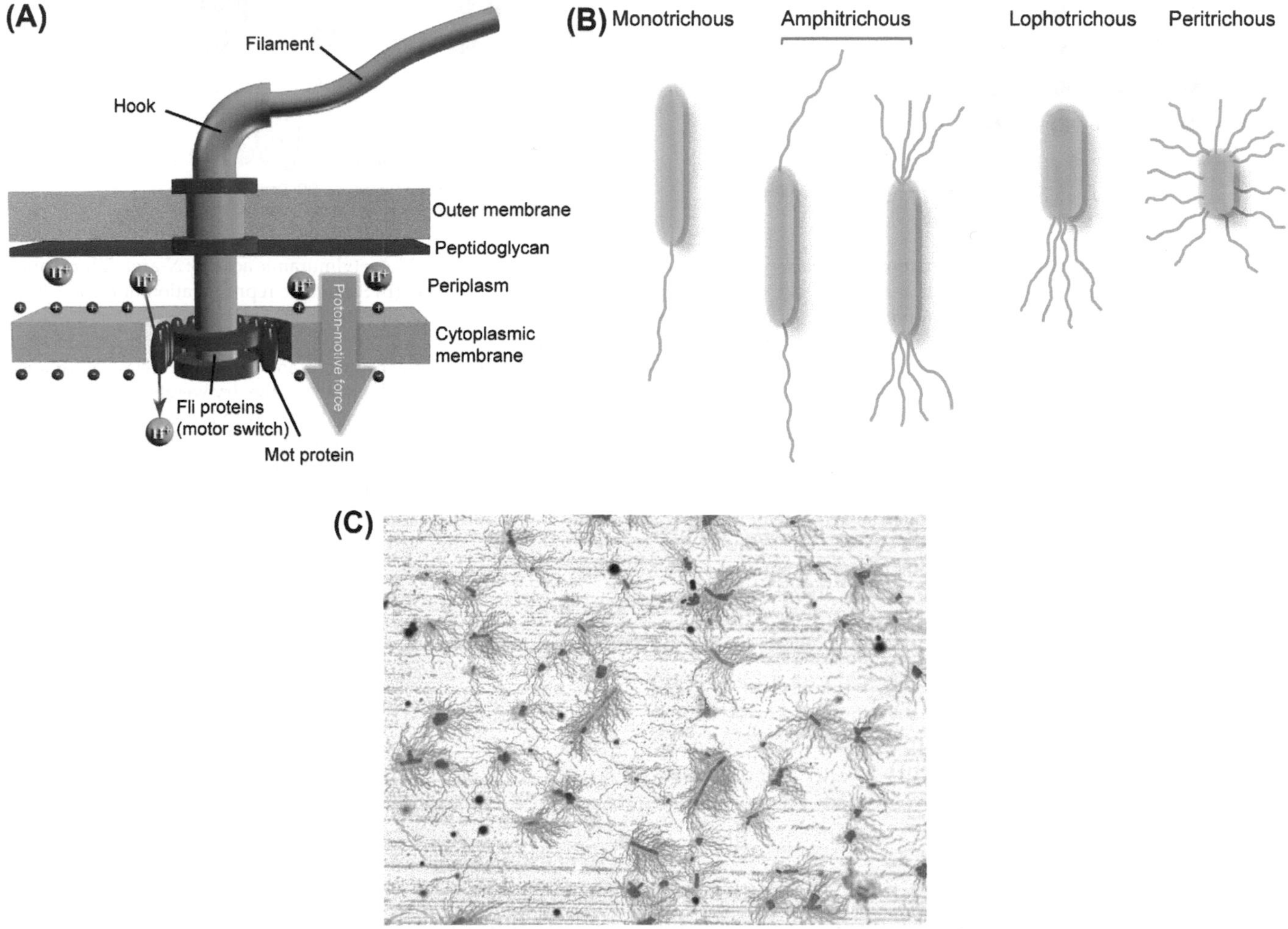

FIGURE 1.7 (A) Schematic of *E. coli* flagellum. Basal body: Located at the base of the flagellum. The basal body, embedded in the cell wall and cell membrane, is the output device. It acts as an engine to provide energy for locomotion. The nearby switch determines the direction of rotation. Hook: This structure points directly away from the cell and has a sharp bend (about 90°) from which filaments protrude. Filament: This filiform structure protrudes from the bacterial cell. It is a hollow tube made of the protein flagellin. Its acts like a ship's or plane's propeller to move the bacterial cell. (B) Examples of bacterial flagellar arrangement. (C) Periplasmic flagella (flagella staining). The bacterial cell is stained red and the flagella are stained light red around the bacterial cell.

pilus or sex pilus according to their morphology, distribution, and function.

The pilus is found on the surface of many gram-positive bacteria and some gram-negative bacteria. It is thinner and shorter than the flagellum. Ordinary pili are 0.3–1.0 μm in length and about 7 nm in diameter and are distributed all over the bacterial cell surface. Sex pili can be found in a handful of gram-negative bacteria. These pili are longer and thicker than ordinary pili, and each bacterial cell can have from one to four sex pili.

1.3.2.4 Spore

The spore is a small round or oval body that forms in bacteria due to cytoplasmic dehydration under unfavorable conditions (Figure 1.8(A)). It is surrounded by multiple membrane layers and has low permeability. Only gram-positive bacteria can form spores, including species such as *Bacillus subtilis*, *Clostridium tetani* (Figure 1.8(B)), etc. The spore contains a complete karyoplasm and enzymatic system and can maintain all the essential activities for the bacteria to remain alive.

The multiple membrane layers of the spore are, from the exterior to the interior, as follows: spore coating, spore shell, outer membrane, cortex, cell wall of spore, and inner membrane, which surrounds the nucleus of the spore.

Spores are difficult to stain due to their thick cell wall. Special staining is required to stain the spore and distinguish it from the bacterial cell (Figure 1.8(B)).

The size, morphology, and location of the spore differ between bacterial species and can be used to help identify bacteria (Figure 1.8(C)). For example, the *C. tetani* spore is round and larger than the transverse diameter of the bacterial cell, forming a drumstick-like structure, as the spore is located at the tip of the bacterial cell (Figure 1.8(B)).

1.3.3 Basic Structure of Virus

Viruses are a kind of acellular microbe consisting mainly of nucleic acid and proteins. Some viruses are composed of a small amount of lipids and polysaccharides. The basic structure of viruses is made up by the

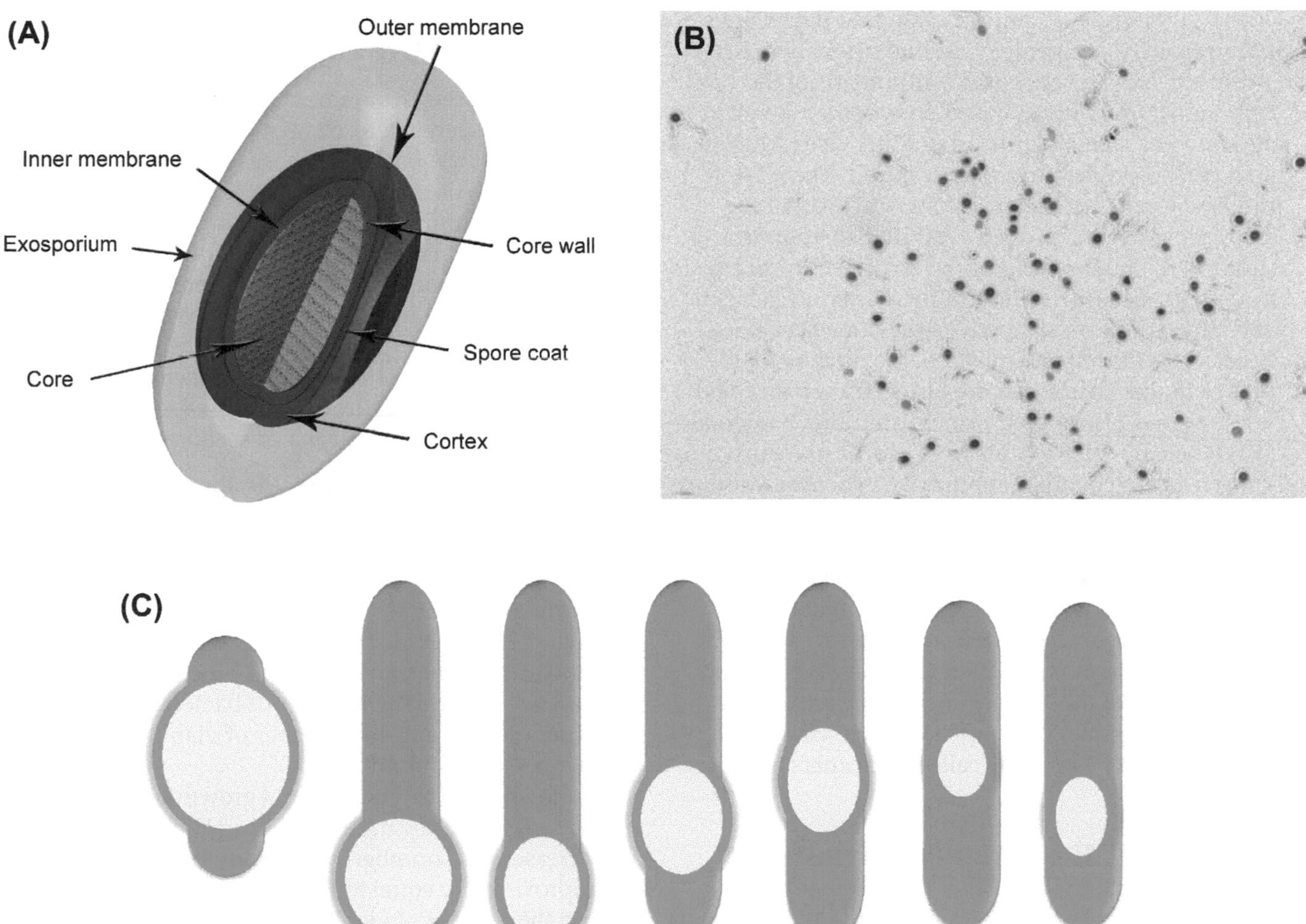

FIGURE 1.8 (A) Schematic representation of the bacterial spore. (B) Spore stain of *C. tetani* (fuchsin-methylene blue stain). (C) Size, morphology, and location of bacterial spores.

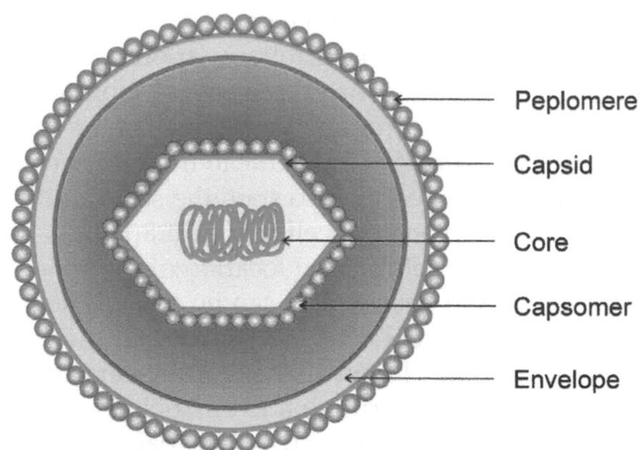

FIGURE 1.9 Basic structure of a virus.

viral core, viral capsid, as well as a membrane envelope in some viruses (Figure 1.9). The size, morphology, and structure of viruses play important roles in viral taxonomy and in diagnosing viral infections.

1. Viral core: namely the nucleic acid component, which makes up the genome of the virus. The viral core provides genetic information that determines pathogenicity, antigenicity, proliferation, heredity, variation, etc. The chemical components of the viral core are DNA or RNA, based on whether the virus is classified as a DNA virus or an RNA virus. Nucleic acid can be single or double stranded. The relative molecular mass of the viral core is $2–160 \times 10^6$.
2. Viral capsid: it is a protein shell that surrounds and protects the nucleic acid of the virus. The viral capsid is sometimes associated with the viral nucleic acid, and this structure is known as the nucleocapsid. In virions without an envelope, the nucleocapsid makes up the entirety of the virus. The viral capsid is composed of repeated protein subunits known as capsomeres, which are made of one or more proteins known as the chemical subunit or structural subunit.
3. Envelope: this is the one or two layers of membrane that surround the capsid of some viruses, which is a structure unique to the class of viruses known as enveloped viruses. The envelope is formed during the maturation process when certain viruses bud out from the cell membrane. Therefore, the envelope can be composed of the host cell membrane and/or the nuclear membrane. The surface of some viral envelopes carries protein protrusions called peplomers or spikes.

1.4 MICROBIAL PHYSIOLOGY

Bacterial cells are synthesis machines that multiply themselves. The growth and division of bacteria include approximately 2000 different types of biochemical reactions that mediate energy conversion or enzymatic biosynthesis.

1.4.1 Binary Fission Reproduction

Binary fission is a process by which many prokaryotes reproduce from a single cell into two new cells (Figure 1.10(A)). Bacteria reproduce asexually by binary fission. Cocci can divide from different planes to form different arrangements. Bacilli divide along their horizontal axis; however, some bacterial species such as *Mycobacterium tuberculosis* occasionally spilt by branching.

During cell division, the cell volume increases and a diaphragm is generated where the cell division is to take place. Then, a single cell will divide into two cells (Figure 1.10(B) and (C)). Under suitable conditions, the majority of bacteria divide quickly, about 20–30 min for one division. However, the growth of some bacterial species can be relatively slow. For example, *M. tuberculosis* takes 18–20 h to complete one round of division.

1.4.2 Bacterial Growth

Mastering the fundamentals of bacterial growth allows the researcher to change culture conditions artificially, adjust the bacterial phases of growth and reproduction, and use beneficial bacteria more efficiently. Bacterial growth involves inoculating a certain number of bacteria into a suitable liquid medium and checking the number of viable cells at different time intervals. With the collected information, it is possible to generate a growth curve using culture time as the horizontal axis and the logarithmic number of viable cells in the culture as the vertical axis (Figure 1.11). Growth curves can generally be divided into four major sections.

1. Lag phase: During this process, the bacteria are adapting to their new environment. The volume of bacterial cells increases and their metabolism is active, but cell division is slow and reproduction is minimal.
2. Logarithmic phase: In this period, bacteria grow rapidly and divide and reproduce at a constant speed. The number of bacteria increases exponentially and the number of viable cells increases logarithmically.
3. Stationary phase: The bacterial growth rate gradually decreases, and the number of dead bacteria increases. The number of newly produced bacteria is approximately equal to the number of dying bacteria, and the number of viable cells remains relatively stable.
4. Decline phase: Bacterial growth rate slows and stops, and the number of dead bacteria is higher than that

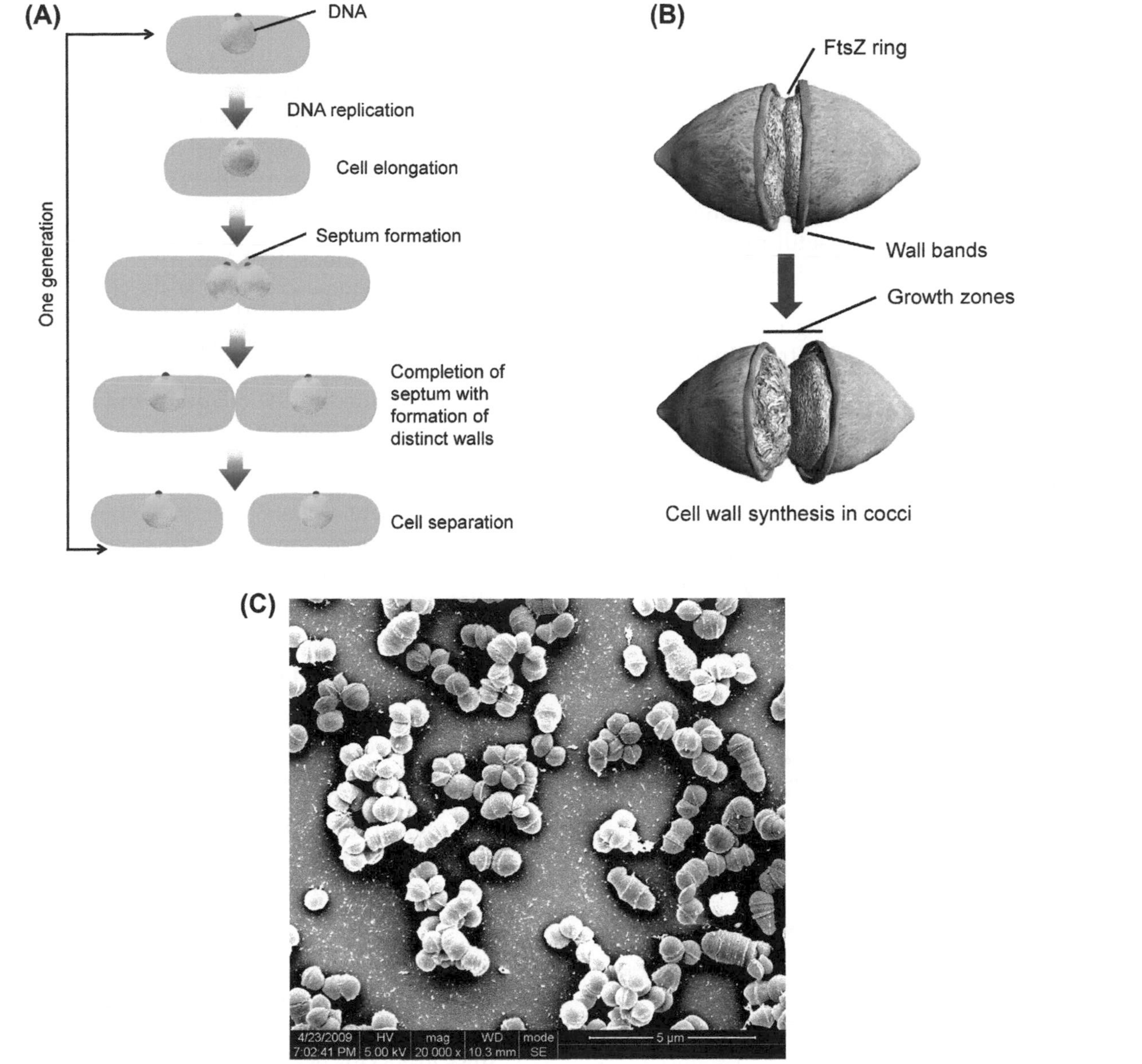

FIGURE 1.10 (A) Binary fission in bacilli. (B) Synthesis of gram-positive bacterial cell wall. During cell division, the volume of the cell increases and a new cell wall is formed. New cell wall materials are added to the preexisting cell wall to maintain structural integrity. (C) Division of gram-positive bacteria (SEM). Cell division in *Streptococcus gordonii*, showing a clear division.

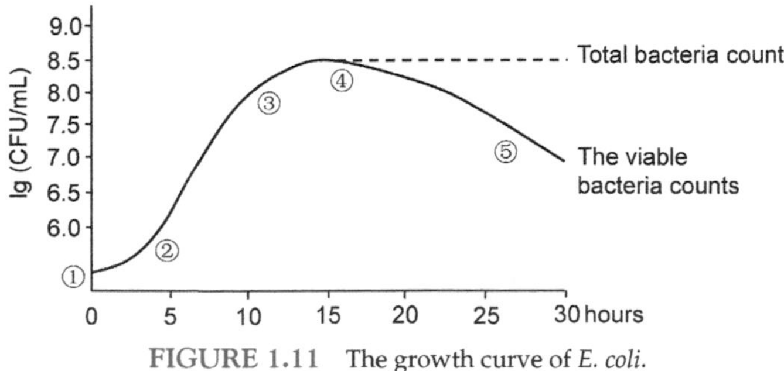

FIGURE 1.11 The growth curve of *E. coli*.

of viable cells. Cells become polymorphic, showing morphologies such as cell deformation, swelling, or autolysis.

1.5 MICROBIAL GENETICS

Like the other living organisms, microorganisms have heredity and variable characteristics. Heredity keeps microbial genetic traits relatively stable to ensure the reproduction of the species, while variations produce changes in the microorganisms that are useful for microbial survival and evolution and ultimately lead to the generation of new species.

1.5.1 Heredity

Heredity is the similarity in biological traits between offspring and its parent. Cells can be considered as a chemical plant in which information can be stored and transformed into a useful product. Enzymes are the molecular machines that catalyze specific chemical reactions. The information is stored DNA, which exists in the cell as two long molecular coiled chains (DNA double helix). The intracellular genetic information is replicated, transcribed, and translated by enzymes, which then leads to protein synthesis (Figure 1.12).

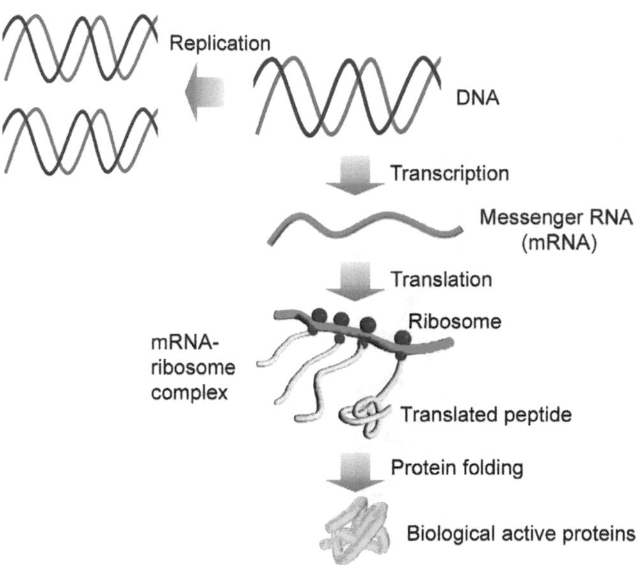

FIGURE 1.12 The processes of replication, transcription, translation of intracellular genetic information and protein synthesis.

1.5.2 Variation

Variation refers to the differences between offspring and its parent under certain conditions, including variations in morphology and structure, virulence, drug resistance, and so on. The variability of microorganisms is divided into genetic variation and nongenetic variation. The former is due to changes in bacterial gene structure. The new characteristics can be stably transmitted to future generations, which is why this type of variation is called genotype variation, as the change is mostly irreversible (Figure 1.13). The latter is caused by the influence of certain environmental conditions that do not change the genetic structure of bacteria. The change is not transmitted to the offspring and is therefore called a phenotypic variation. Genetic variation is rarely influenced by external environmental factors. Therefore, genetic variation tends to occur in individual bacterial cells, while phenotypic variation tends to occur in a bacterial flora due to the effect of environmental factors. These variations can revert with the removal of the stimulating environmental factors.

1.5.3 Genetic Material of Bacteria

Nucleic acids are the basis of organismal heredity. Two types of nucleic acid exist: DNA and RNA. DNA is the genetic material in prokaryotic and eukaryotic organisms, while the genetic material in viruses is DNA and RNA. The genetic material found in microorganisms includes chromosomes, plasmids, bacteriophages, and transposable elements.

Bacteriophages are viruses that infect bacteria, fungi, actinomycetes, mycoplasmas, and spirochetes. They inject their genetic material into the infected host cell and can induce bacterial cell lysis under certain conditions. Bacteriophages, known as phages, can only reproduce in specific host strains and have high specificity for

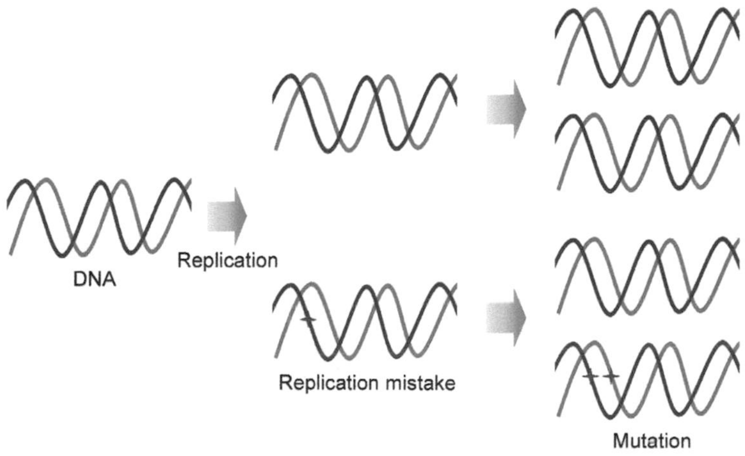

FIGURE 1.13 DNA mutation.

their host. The specificity is related to phage cell binding molecules and the structure and complementarity of the host bacterial strain's surface receptor molecules.

Since phages are small, they can only be observed under electron microscope. They can be divided into three basic morphologies: tadpole-shaped, spherical, and rod-shaped. Most phages are tadpole-shaped and consist of a head and a tail (Figure 1.14(A)).

The relationship between the phage life cycle and its host bacteria is shown in Figure 1.14(B). Phages that infect bacteria can produce two outcomes. The first, observed in virulent phage, involves phage multiplication resulting in the production of many progeny phages, bacterial cell lysis, and cell death. This cycle is known as the lytic cycle. The second outcome is lysogeny, observed in temperate phages. It involves the integration of phage nucleic acid with the bacterial chromosome, resulting in the formation of a prophage. The phage genetic material is reproduced when the bacterial cell divides.

1.5.4 Mechanism of Microbial Variation

Genetic variation in microorganisms has its basis in mutations caused by changes in their genetic sequence. These changes are stable and heritable. Mutations can generally be classified as gene mutations and chromosomal aberrations. The spontaneity and randomness of microbial mutations can be tested by using fluctuation test or replica plating (Figure 1.15(A) and (B)).

1.5.5 Gene Transfer and Recombination

Gene transfer is a process by which exogenous genetic material from a donor cell is transferred to the receptor cell. However, simply the process of transferring genetic material is not enough, as the recipient cell must be able to accommodate exogenous genes. Integration between the transferred gene and the DNA of the recipient cell is a process known as recombination, whereby the recipient cell acquires certain characteristics of the donor strain. Gene transfer and recombination in bacteria can take place through processes such as transformation, conjugation, transduction, cell fusion, and lysogenic conversion.

1.5.5.1 Transformation

Transformation takes place when donor bacteria DNA is cleaved and free DNA fragments are directly taken up by receptors. As a result, recipient cells acquire certain genetic traits from the donor cell. This phenomenon was confirmed in *Streptococcus pneumoniae*, *Staphylococci*, and *Haemophilus influenzae*. Griffith first showed that bacterial transformation takes place by infecting mice with *S. pneumoniae* in 1928. An outline of the experiment is shown in Figure 1.16.

1.5.5.2 Conjugation

Conjugation is the method by which bacteria physically connect with one another through their pilus to transfer genetic material (mainly plasmid DNA). Plasmid transfer from the donor to the recipient cell results in the recipient cell acquiring some of the genetic traits of the donor cell. Plasmids that can be transferred through conjugation are called conjugative plasmids, which include the F, R, Col, and virulence plasmids. The F plasmid encodes the pilus, controls pilus formation and whether or not the pilus enables conjugation (Figure 1.17(A)).

Plasmids that cannot be transferred between bacteria through a pilus are called nonconjugative plasmids.

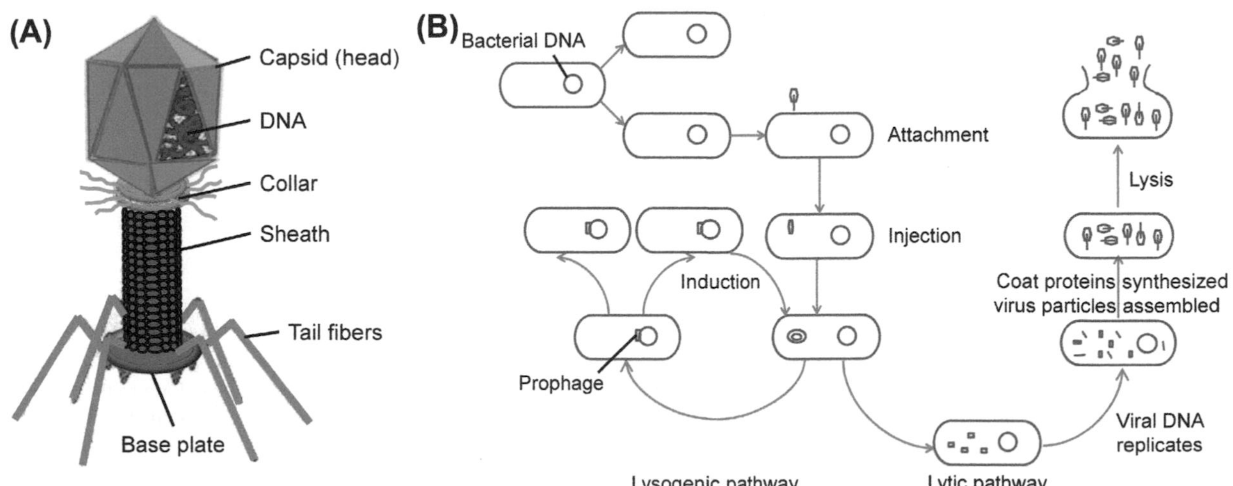

FIGURE 1.14 (A) The structural model of a phage. The phage head is icosahedral and is approximately 80 × 100 nm in size. It consists of the capsid protein that surrounds the internal nucleic acid. The tail is a tubular structure composed of protein, including the tail whiskers, tail collar, tail tube, tail sheath, tail fibers, and tail baseplate. (B) Lysogenic and lytic cycles of lysogenic phage.

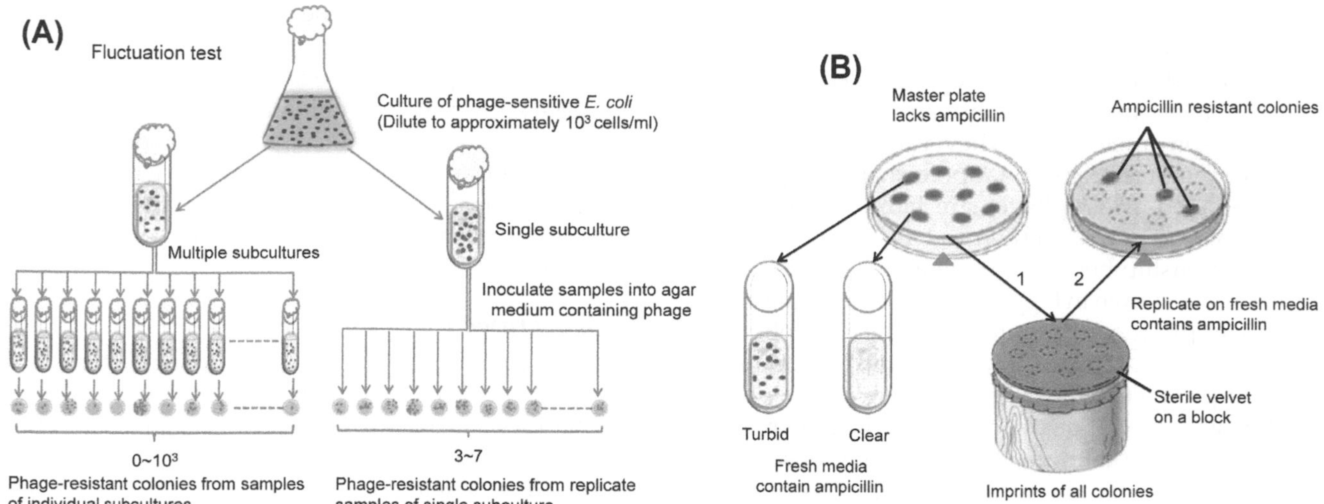

FIGURE 1.15 (A) Fluctuation test: The fluctuation test shows that mutations preexist in a population of bacteria in the absence of selection. This was tested by Luria and Delbruck using naturally existing phage-resistant mutations in bacterial populations. A given concentration (10^3/ml) of *Escherichia coli* sensitive to specific phages was inoculated in two equal volumes (10 ml) of broth medium. One inoculum was concentrated in a large test tube and the other was evenly distributed into 50 small test tubes. After 24–36 h incubation under the same conditions, the bacterial cultures from the large and small tubes were plated onto phage-containing plates, and the number of colonies was measured. The results typically showed that of 50 phage-coated plates inoculated with bacteria from the large test tube, the fluctuation in the colony number was small (3–7). On the other hand, when the 50 small tubes were plated onto 50 plates, the number of the colonies tended to fluctuate significantly, from zero colonies to several hundreds. (B) Replica plating: Replica plating (Lederberg et al., 1952) involves plating antibiotic-susceptible strains on agar plates in the absence of antibiotics until scattered single colonies grow. Using a block covered with sterile velvet, gently press the velvet onto the surface of the agar plates so that bacterial colonies are imprinted onto the sticky velvet surface. Then, press the velvet surface onto an agar plate containing antibiotics. After an appropriate culture time, bacteria susceptible to the antibiotics are completely suppressed, but drug-resistant colonies will be visible on the plate. We can find colonies corresponding to drug-resistant colonies on the original plates lacking antibiotics. The drug-resistant colonies can then be transplanted to culture broth containing the appropriate antibiotics to observe bacterial growth. Although the bacterial colonies on the original agar plate have never been in contact with antibiotics, they are nonetheless resistant to antibiotic drugs.

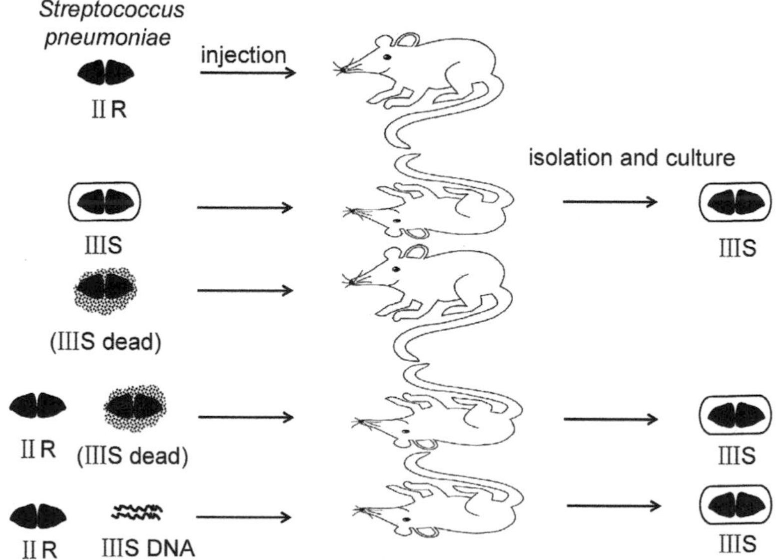

FIGURE 1.16 ***Streptococcus pneumoniae*** **transformation in mice.** *S. pneumoniae* with a polysaccharide capsule belong to the virulent type III strain. These colonies are smooth (S) in appearance. *S. pneumoniae* without the polysaccharide capsule belong to the avirulent type II strain and appear as rough (R) type colonies. In the classic Griffith experiment, type II-R bacteria and type III-S bacteria were injected into mice. Mice that received the type II-R bacteria survived, and those that received the type III-S bacteria died. The type III-S bacteria were isolated from the blood of the dead mice and were heated until they were no longer active. These dead type III-S bacteria were injected into mice, and the mice survived. However, when dead type III-S bacteria and live type II-R bacteria were both injected into the same mice, the mice died, and type III-S bacteria were isolated from their blood. This experiment showed that the live type II R-type bacteria were able to obtain genetic material from dead type III-S bacteria that transformed them from an avirulent strain to a virulent one. It also suggests that the genetic material encoded the capsule virulence factor from type III-S bacteria.

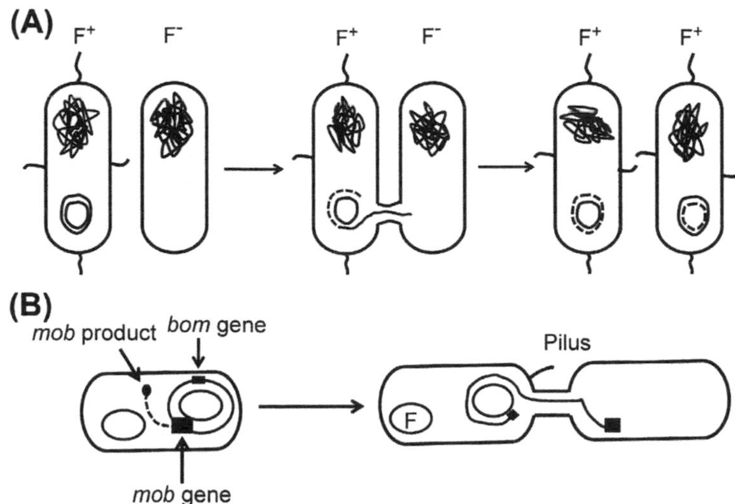

FIGURE 1.17 (A) The transfer and replication of the F plasmid during conjugation. Bacteria possessing the F plasmid and F-pili are the male equivalent strain (F^+), while bacteria lacking the F plasmid and F-pili are the female equivalent strain (F^-). When the F^+ and F^- strains are present in the same environment, the F-pilus from the F^+ bacterial cell conjugates to the F^- surface receptor. The F-pilus gradually shortens so that the two cells are pulled close to each other and a channel is formed. Plasmid DNA from the F^+ bacteria breaks at the origin of transfer (*oriT*) and the 5′ end extends through the channel into the F^- bacteria. Single-stranded DNA in both bacterial cells replicate by rolling circle replication, and each cell forms a complete F plasmid. Therefore, the donor cell has not lost its F plasmid after the transfer, and the recipient cell becomes F^+ strain bacteria after receiving the F plasmid. (B) Nonconjugative plasmid (ColE1) induced to transfer by F plasmid. The ColE1 plasmid can be induced to transfer from a donor to a recipient cell. This process requires two genes encoded on the ColE1 plasmid: the specific site *bom* on ColE1 DNA (also known as the *nic* locus) and the nuclease encoded by the *mobA* gene. When both F and ColE1 plasmids are present in the same bacterial cell, the *mobA* gene is transcribed and its product creates a single strand break at the *bom* locus to form a gap. As a result, the ColE1 plasmid goes from a supercoiled plasmid to an open loop.

The nonconjugative plasmid ColE1 is relatively low in molecular mass and does not encode the necessary gene required for it to be transferred from one cell to another. However, if there is another conjugative plasmid in the host cell, ColE1 can tag along and be transferred from one cell to another. For example, the F plasmid that encodes the genes necessary for pilus formation can help ColE1 transfer from one cell to another (Figure 1.17(B)).

Conjugation is widespread in gram-negative bacteria and can be observed in almost all members of the *Enterobacteriaceae*. Some gram-positive bacteria have been reported to conjugate (e.g., *Streptococcus*, *Bacillus subtilis*), and the phenomenon has also been observed in *Streptomyces*.

1.5.5.3 *Transduction*

Transduction uses a temperate phage as the vehicle by which DNA from a donor cell is transferred into a recipient cell. Transduction can transfer larger fragments of DNA than transformation. According to the genes involved in transduction, the process can be divided into generalized transduction and restricted transduction (Figure 1.18(A) and (B)).

Generalized transduction can transfer plasmids. The transduction of plasmid R in *S. aureus* is a very important clinical feature.

In generalized transduction, the packaged DNA can be from any part of the donor strain chromosome. When phages exit the lysogenic phase, the prophage will excise

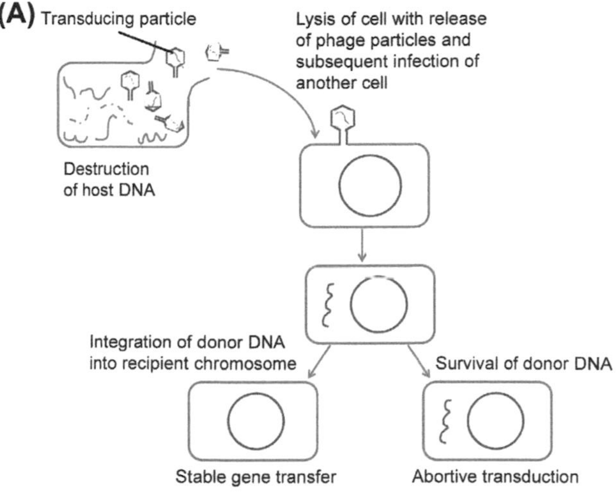

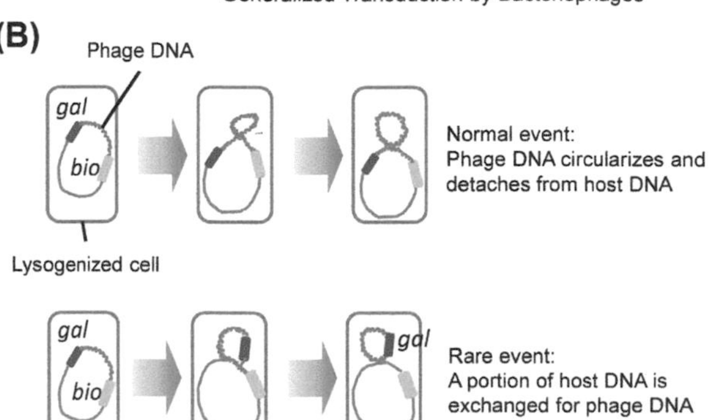

FIGURE 1.18 (A) Generalized mode of transduction. (B) Restricted mode of transduction.

itself from the bacterial chromosome and enter into the lytic phase. In the latest stages of the lytic phase, phage DNA is replicated in large quantities and errors can occur during phage assembly. Approximately one in 10^5–10^7 phages will contain bacterial DNA fragments that have been mistakenly packaged into the phage head, which results in the formation of a transducting phage. Transducting phages can infect another host bacterium and inject the DNA fragment they are carrying into the recipient cell, thereby transferring DNA from one bacterial cell to another.

Restricted transduction, or specific transduction, describes the process in which transduction is restricted to specific genes from the chromosome of the donor strain. If phage λ is transferred into *E. coli* K12 while it is in the lysogenic phase, the phage DNA is integrated into a specific site in the *E. coli* chromosome, between the galactose (*gal*) and biotin (*bio*) genes.

CHAPTER

2

Techniques for Oral Microbiology

2.1 SMEAR AND STAIN TECHNIQUES

Smear test and slide stain are basic techniques for microbial identification and are primarily used for morphological observation. The combination of smear and stain is widely used in oral microbiology research to differentiate between spirochetes, bacteria, fungi, and protozoa, as well as to identify specific cellular structures including spore, capsule, flagellum, and others. Currently, some of the commonly used techniques for cellular morphology examination are direct smear test and stained smear test, including Gram stain and Congo red stain. The procedure for smear test is shown in Figure 2.1.

Specific stains can be used to observe the specific microbial structures, including the bacterial spore, capsule, flagellum, fungal hypha, clamydospore, etc.

2.1.1 Kopeloff's Modification of Gram Stain

Gram stain is a commonly used method for the identification of bacteria and fungi. After Gram staining, bacterial species can be generally differentiated into gram-positive and gram-negative groups, and most clinical anaerobic microbes can be differentiated by their typical cellular morphology. In addition, *Mycoplasma* with their ringshape can also be identified. Kopeloff's modification of Gram stain is recommended by the Virginia Polytechnic Institute (VPI) for better visualization and differentiation of microbes.

2.1.1.1 Mechanism

Many theories have been proposed to explain the mechanism of Gram staining, including isoelectric point theory, chemical theory, and permeation theory. Among them, permeation theory has gained wide acceptance. It is believed that crystal violet placed on the microbe smear slides during staining interacts with aqueous iodine to form insoluble precipitates within the cell. During decolorization, alcohol and acetone can dissolve the lipid in the outer membrane and leach the precipitated dye–iodine complex out of the microbial cell. Gram-positive microbes have a thick cell wall with highly cross-linked peptidoglycans and relatively fewer lipids. Thus, gram-positive microbes cannot be easily decolorized and become deeply stained in purple. In contrast, gram-negative microbes have a thin membrane with high lipid content and little peptidoglycan. The precipitated dye–iodine complex can therefore be easily dissolved and leave the cell. As a result, the cell is stained red from the counterstain.

2.1.1.2 Preparation of Gram Stain Reagents (Commercially Available)

1. Crystal violet

Liquid A (add 10 g crystal violet to 1000 ml distilled water and mix well).

Liquid B (add 50 g $NaHCO_3$ to 1000 ml distilled water and mix well).

2. Gram's iodine

Dissolve 4 g NaOH into 25 ml distilled water and add 20 g iodine and 1 g KI. After everything is completely dissolved, add 975 ml distilled water, mix well, and store the liquid in a brown bottle.

3. Decolorizers

Add 300 ml acetone to 700 ml 95% alcohol and mix well.

4. Fuchsin counterstain

Add 5–10 g basic fuchsin to 100 ml 95% alcohol to form a supersaturated liquid. Mix 10 ml of the supersaturated liquid with 90 ml 5% carbolic acid solution followed by the addition of 900 ml water. Mix well and filter through filter paper. (Safranin solution can also be used in place of fuchsin solution for counterstain: add 20 g safranin to 100 ml 95% alcohol and adjust the volume to 1000 ml with distilled water.)

Atlas of Oral Microbiology
http://dx.doi.org/10.1016/B978-0-12-802234-4.00002-1

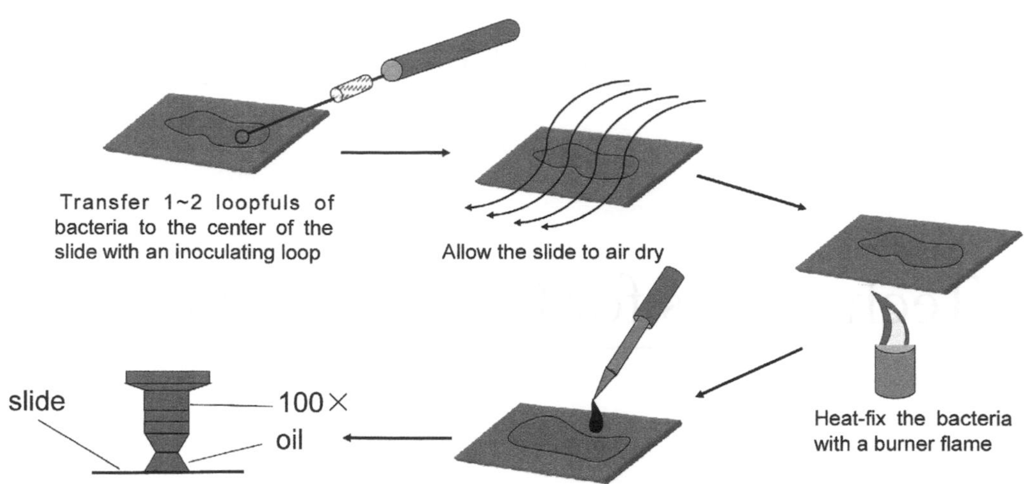

FIGURE 2.1 Procedure of smear test.

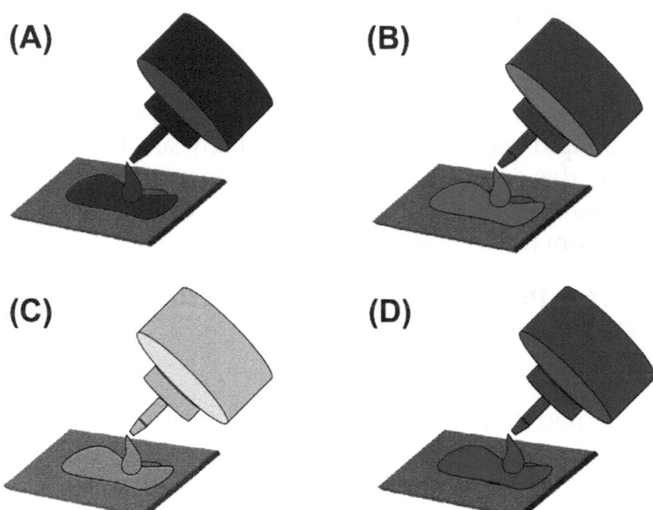

FIGURE 2.2 Gram stain procedure.

2.1.1.3 *Staining Procedure*

Gram stain procedure is shown in Figure 2.2.

1. Add crystal violet Liquid A onto the fixed smear slide. After 5 s, add Liquid B and gently shake the slide to mix Liquids A and B for 30 s. Gently rinse off the crystal violet with tap water.
2. Flood the smear slide with Gram's iodine for 30 s. Gently rinse off the iodine with tap water.
3. Add decolorizer to the smear while holding the slide at an angle to allow the decolorizer to drain. Gently shake the slide for 5–10 s until crystal violet leaches out. Gently rinse off the decolorizer with tap water.
4. Flood the smear with fuchsin or safranin counterstain for 5–10 s. Gently rinse off any excess fuchsin or safranin with tap water. Drain and air dry the slide before observation.

2.1.1.4 *Reporting Results*

Examine the slide under a light microscope using oil immersion. Gram-positive bacteria appear purple (Figure 2.3(A) and (B)), while gram-negative bacteria appear pinkish red (Figure 2.3(C) and (D)).

2.1.1.5 *Precautions*

1. Smear only a monolayer of cells on the slide.
2. Avoid overheating of the smear slide.
3. Slightly prolong the time of decolorization if the smear layer is thick.
4. Old cultures of gram-positive bacteria may appear as gram-negative bacteria.

2.1.2 Staining Bacterial Spores

Gram stain and fuchsin-methylene blue are commonly used to observe bacterial spores.

2.1.2.1 *Gram Stain*

The Gram stain procedure is the same as described above. The result of Gram stain is shown in Figure 2.4.

2.1.2.2 *Fuchsin-Methylene Blue Stain*

2.1.2.2.1 Reagent Preparation

1. Fuchsin solution: Add 5–10 g basic fuchsin to 100 ml 95% alcohol to form a supersaturated liquid. Mix 10 ml of the supersaturated liquid with 90 ml 5% carbolic acid solution followed by 900 ml water.
2. Methylene blue solution:

   Liquid A: Add 0.3 g methylene blue to 30 ml 95% alcohol.
   Liquid B: Add 0.01 g KOH to 100 ml distilled water.
   Mix Liquids A and B.

3. Decolorizer: 95% alcohol

Stain procedure: Flood the smear slide with fuchsin solution. Heat the slide on the alcohol burner to steam for 3–5 min (avoid drying out the smear slide and add fuchsin solution when necessary). Cool off the slide and rinse off excess fuchsin under tap water. Flood the slide with decolorizer for 1 min and rinse. Stain

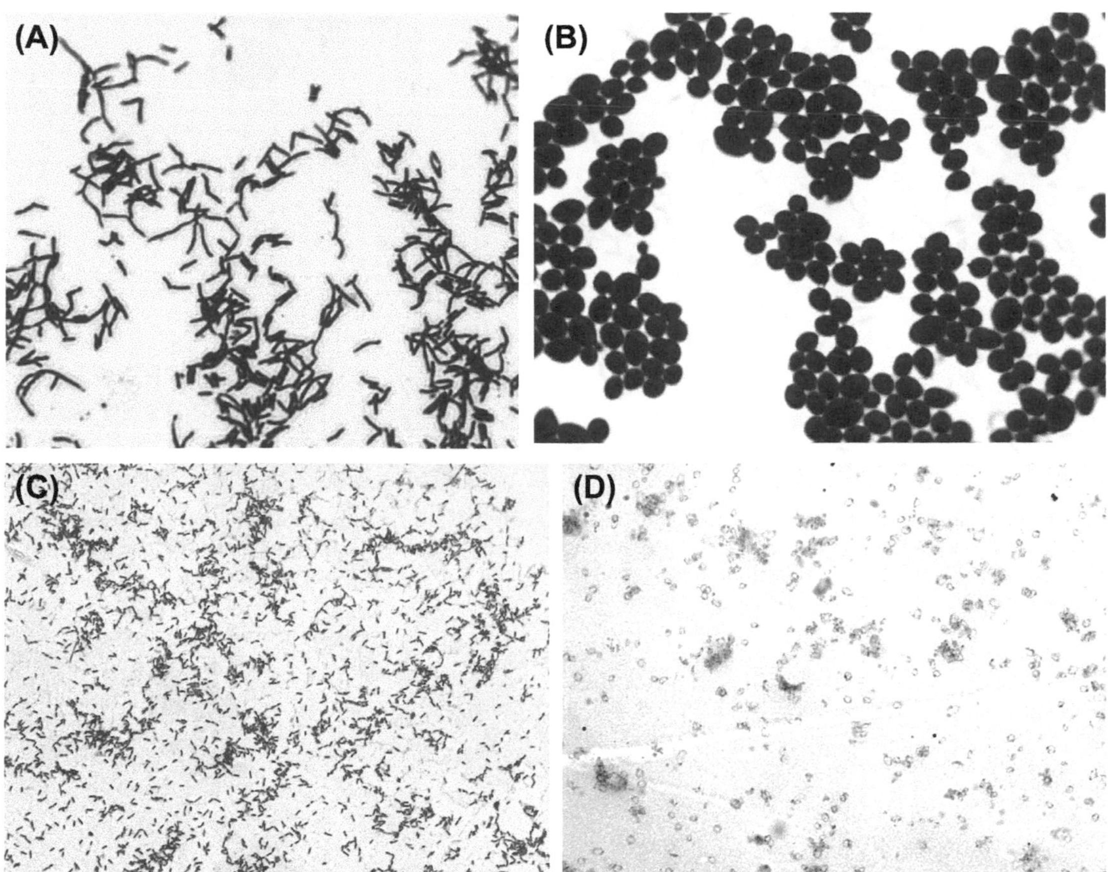

FIGURE 2.3 (A) Gram-positive bacteria (*Actinomyces naeslundii*). (B) Gram-positive yeast (*Candida albicans*). (C) Gram-negative bacteria (*Eikenella corrodens*). (D) Gram-negative bacteria (*Mycoplasmal pneumonia*).

the slide with methylene blue solution for 0.5 min and rinse. Observe the slide under a microscope using oil immersion. Results of spore staining are illustrated in Figure 2.5.

2.1.3 Staining the Bacterial Capsule

The bacterial capsule can be stained using a variety of methods, including the Murs stain, ink methylene blue stain, Hiss copper sulfate stain, Ott safranin stain, etc. Encapsulated bacteria that are found in the oral cavity include *Streptococcus pneumoniae*, *Porphyromonas gingivalis*, and others.

2.1.3.1 Murs Stain

Preparation of staining solutions:

1. Staining solution (carbolfuchsin solution): Add 1 g basic fuchsin to 10 ml 95% ethanol solution, then add 40 ml 5% aqueous carbolic acid solution and mix.
2. Mordant staining solution: Mix 100 ml 2% tannic acid, 250 ml 10% potassalumite ($K_2SO_4 \cdot 12H_2O$), and 100 ml 7% saturated mercuric chloride.
3. Counterstaining solution (alkaline methylene blue solution): Add 0.3 g methylene blue to 30 ml 95% ethanol solution, then add 100 ml 0.01% potassium hydroxide solution.
4. Destaining solution: 95% ethanol.

Staining protocol: Prepare a smear sample using physiological saline and fix. Heated carbol fuchsin solution is used to stain the smear for 1 min. Flush with water after destaining using 95% ethanol. Stain with mordant staining solution for 0.5 min, and destain with 95% ethanol for 0.5 min. Stain with alkaline methylene blue solution for 0.5 min, then flush the stain solution with water. Examine microscopically after air drying. Typical results are shown in Figure 2.6.

2.1.3.2 Wright's Stain

Bacterial capsules can also be stained with the Wright's staining method. Furthermore, the Wright's staining method can be used to stain *Rickettsia* and spirochetes, which appear purple after staining.

Preparation of staining solutions:

1. Wright stain solution: Weigh out 0.1 g dry Wright's stain (same as eosin methylene blue stain). Grind eosin Y using mortar and pestle until it becomes a fine powder. Add 10 ml methanol to the mortar to fully dissolve the eosin Y. Add 20 ml methanol, mix, and allow to stand for a moment. Decant the liquid into a clean storage bottle, add methanol to the mortar, mix, and repeat several times until all of

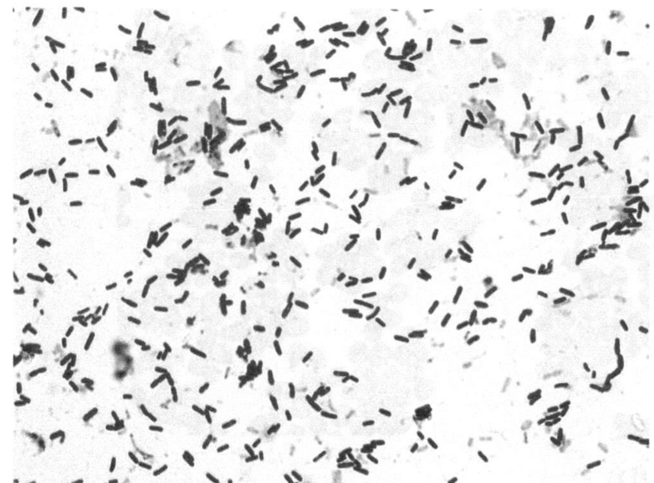

FIGURE 2.4 ***Bacillus subtilis*** **(Gram stain).** Under light microscope with oil immersion, bacterial spores appear colorless with strong refraction. They are located toward the center (central spore) or ends of the cell (terminal spore) with round or ovoid shape.

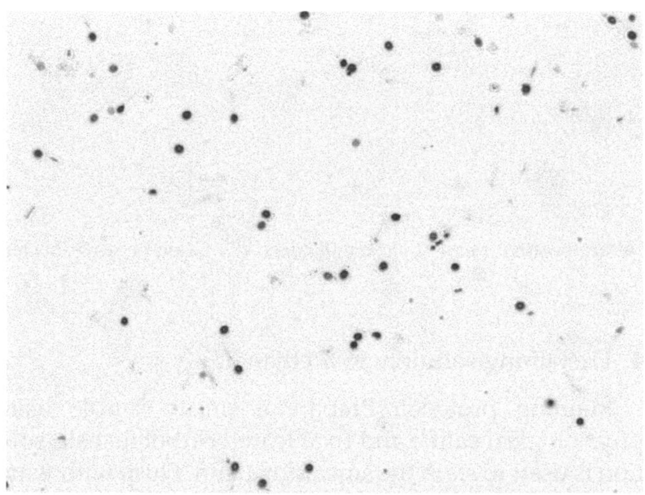

FIGURE 2.5 **Spore stain with *Clostridium tetani* (fuchsin-methylene blue stain).** The cell appears blue and the spore appears red. A round spore is located at the end of the cell, giving the cell a drumstick appearance.

the stain is dissolved. Sixty milliliters of methanol should be used for 0.1 g stain. Shake and seal the storage bottle, and store it in the dark at room temperature.

2. Phosphate buffer (pH 6.4–6.8): 30 ml 1% KH_2PO_4, 73.5 ml M/15 KH_2PO_4, 20 ml 1% Na_2HPO_4, 26.5 ml M/15 Na_2PO_4, complete with H_2O to 1000 ml.

Staining protocol: Prepare a smear using the conventional method, and allow to air dry. Place four to six drops of Wright's stain onto the smear and allow to stain for 1 min. Add an equal amount of phosphate buffer (pH 6.4–6.8). Shake gently to allow the phosphate buffer and the staining solution to mix evenly. Let the stain stand for 6–8 min and flush with water. The smear should

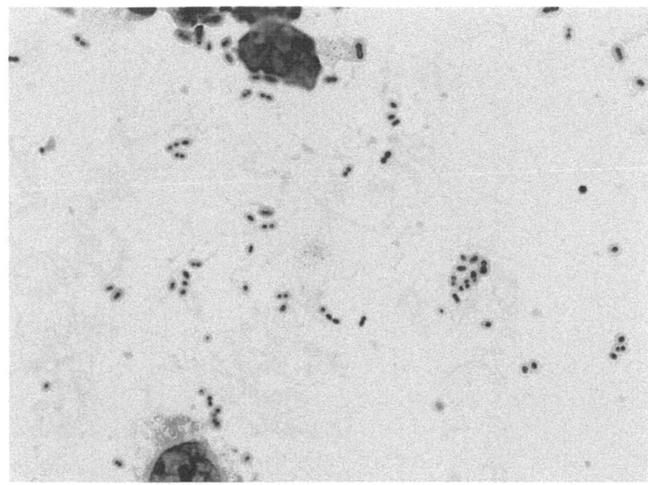

FIGURE 2.6 **Capsule of *Streptococcus pneumoniae* (Murs staining method).** Bacterial cells are stained red, and the capsule around the cell shows up as blue transparent circles.

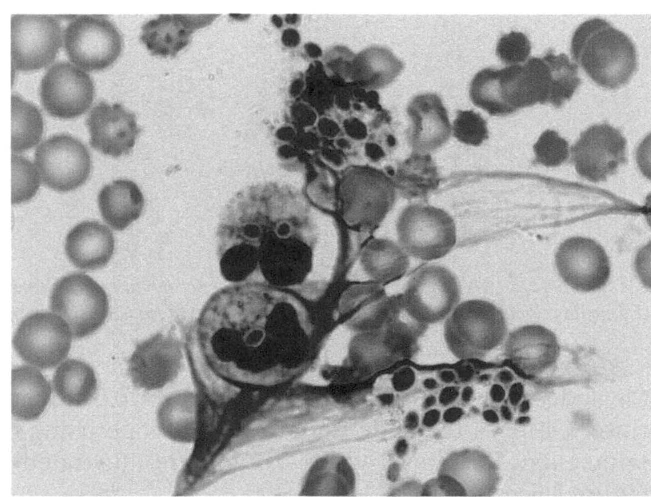

FIGURE 2.7 **Capsules of *Histoplasma capsulatum* (Wright's stain).**

appear pink (if the smear appears bluish violet, it can be destained by using 20% hydrochloric acid methanol solution). A typical result is shown in Figure 2.7.

2.1.4 Flagellar Stain

Flagellar stain is used to observe whether bacteria have flagella, as well as the number and location of flagella. It is a helpful tool for bacterial identification. Flagellar staining is of great importance in identifying motile oral bacteria. For example, *Selenomonas* can have up to 16 flagella, while straight *Campylobacter* only has one polar flagellum.

2.1.4.1 Preparation of Staining Solutions

Liquid A: 20 ml 2% tannic acid solution dissolved in heated water bath, 20 ml 20% potassium solution (heat

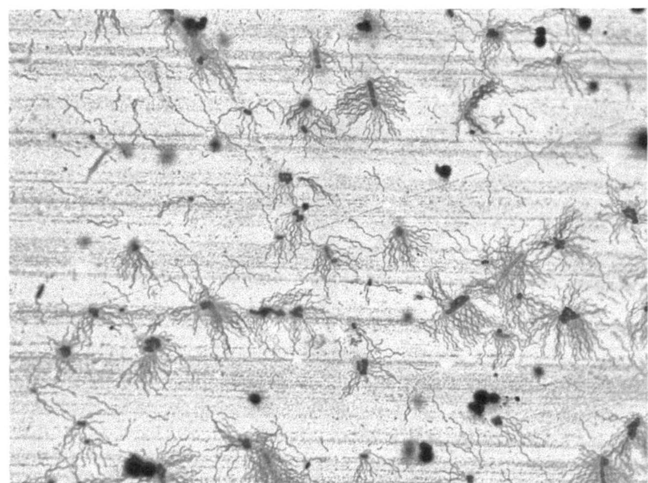

FIGURE 2.8 **Periplasmic flagella (flagella staining).** The bacterial cell is stained red, and the light-red flagella can be seen around the bacterial cell.

to dissolve), and 50 ml 5% aqueous solution of carbolic acid. Mix all three solutions.

Liquid B: alkaline fuchsin ethanol saturated liquid (same as in Gram stain solution). Mix 30 ml Liquid A with 10 ml Liquid B, and filter. Keep the filtrate for 6–10 h at room temperature for optimum results.

2.1.4.2 Smear Preparation

Flagellar staining requires correct preparation of the smear. Pick a small, young agar culture farthest away from the original streak, and resuspend in a small tube in sterile distilled water. Keep the small tube at room temperature for 20–30 min, or keep it in a 37 °C incubator for 4–5 min to allow the culture to become fully resuspended. Remove a loop of the bacterial suspension with a disposable sterile loop, and place it gently on a clean slide. Tilt the slide gently, or push the bacterial suspension with the loop. Then allow the smear to dry naturally at room temperature or in the 37 °C incubator.

2.1.4.3 Staining Protocol

Drop the Liquid A and Liquid B mixture onto a dry smear and allow to stain for 1–2 min. Flush the slide with water. When the smear is dry, observe the result under an oil immersion lens (Figure 2.8).

2.1.5 Staining Procedure for Special Fungal Structures

Gram stain and lactophenol cotton blue stain are used to visualize certain fungal structures that are significant for identification. These include the germinal tube, hyphae, and spores and are especially important in clinical analyses of oral mucosal diseases. For example, in the identification of thrush, angular stomatitis is closely related to *C. albicans*. The checking specimen is originated from albuginea or focal secretion of oral mucosa. Gram staining and acid phenol cotton blue stains are often used to verify the structures in fungi such as germ tubes, hyphae, and spores, especially for oral mucosal diseases such as thrush and angular cheilitis-related *C. albicans*. Most specimens are taken from the oral mucosa lesions albuginea and secretions.

2.1.5.1 Crystal Violet Stain

Crystal violet staining solution is prepared in the same way as Liquid A used in Gram stain. Take a small quantity of culture and mix with physiological saline to prepare a smear. Stain the smear with crystal violet solution. Observe under oil immersion lens (Figure 2.9(A) and (B)).

2.1.5.2 Lactophenol Cotton Blue Stain

Preparation of staining solution: Dissolve 20 g carbolic acid (solid), 20 ml lactic acid, and 40 ml glycerol into 20 ml distilled water (heat as gently as possible). Add 0.05 g cotton blue, shake until well mixed, and filter before storing.

Staining protocol: Place a drop of lactophenol cotton blue staining solution onto a clean slide, and mix the fungal culture or clinical sample with the staining solution. Place a coverslip on top and heat gently. Press the coverslip gently to remove any bubbles. Observe the slide under an oil immersion lens (Figure 2.10(A) and (B)).

Germinal tube and blastospores are stained bright blue after *C. albicans* are incubated in 0.5–1 ml human serum or in sheep serum for 2–4 h at 37 °C and stained with lactophenol cotton blue stain.

Pseudohypha chlamydospores of *C. albicans* are bright blue, and the big, spherical, thick-walled chlamydospore is at the tip or side wall of the pseudohypha.

2.1.6 Giemsa Stain for *Mycoplasma*

Giemsa stain is used to observe *Mycoplasma* (Figure 2.11).

2.1.7 Negative Congo Red Staining of Plaque Bacteria

Due to the simplicity of this method in producing high quality stained smears, the Congo red staining method has been widely used in the examination of periodontitis and other oral clinical specimens.

1. Sample: Saliva, plaque, other oral specimens.
2. Staining solution: 2% Congo red solution.
3. Staining procedure: Place a drop of the 2% Congo red solution on a clean slide. Mix the specimen and the staining solution, and spread the mixture thinly with a slide. After the smear dries naturally, smoke it

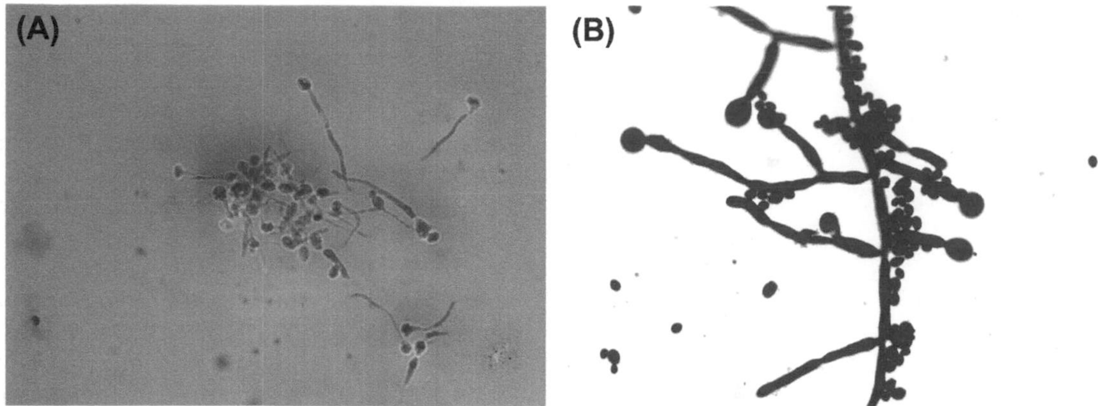

FIGURE 2.9 (A) Germinal tube and blastospore of *Candida albicans* (crystal violet stain). The germinal tube and blastospore appear purple when *C. albicans* is incubated in 0.5–1 ml human serum or in sheep serum for 2–4 h at 37 °C and stained by the crystal violet staining method. (B) Pseudohypha chlamydospore of *C. albicans* (crystal violet stain). Pseudohypha chlamydospores of *C. albicans* appear purple when stained with crystal violet. The chlamydospores are big, spherical, thick-walled, and they are located at the tip or side wall of unevenly stained pseudohypha.

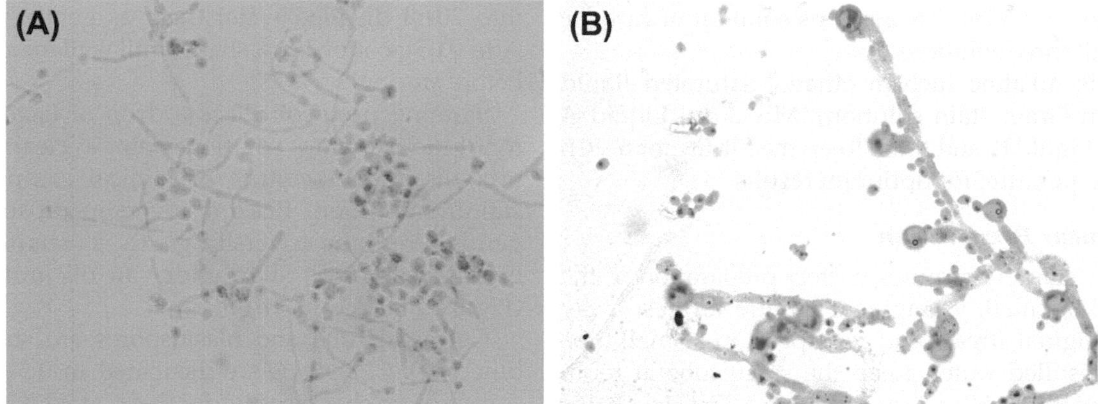

FIGURE 2.10 (A) Germinal tube and blastospore of *C. albicans* (lactophenol cotton blue stain). (B) Pseudohypha chlamydospore of *C. albicans* (lactophenol cotton blue stain).

over a bottle of concentrated hydrochloric acid until the red smear turns blue.

4. Results: The blue smear is examined under light microscope under oil immersion lens (Figure 2.12).

Listgarten classification is generally used in negative Congo red staining and dark-field microscopy to classify the observed plaque bacteria. The method involves selecting an evenly spread field, counting 200 bacterial cells, and reporting the percentage of different species according to their morphology.

Coccoid cells: Cell diameter of 0.5–1.0 μm, including several kinds of coccobacilli.
Straight rods: Cells measure approximately 0.5–1.5 μm in width, 1.0–1.9 μm in length. Some types of mycobacteria are included.
Filaments: Cells measure 0.5–1.5 μm in width, and the ratio of length to width is greater than 6. Most bacteria are shaped as irregular long filamentous cells.
Fusiform cells: Cells measure approximately 0.3–1.0 μm in diameter, 10 μm in length, and show tapered ends.
Curved rods: Cells are similar in dimension to straight rods, but are curved or crescent-shaped.
Spirochetes: Cells have a spiral shape and measure 0.2–0.5 μm in width and 10–20 μm in length.

2.1.8 Protozoan Smears

Entamoeba gingivalis and *Trichomonas tenax* are the main protozoans found in the oral cavity. Wet smears can be prepared using fresh specimens to observe the morphology and mobility of the protozoans. Giemsa stain can be used on a fixed specimen to observe the cellular structure.

2.1.8.1 Wet Smear for Fresh Samples

Mix gingival margin plaque samples or subgingival plaque with saline to prepare fresh wet smears.

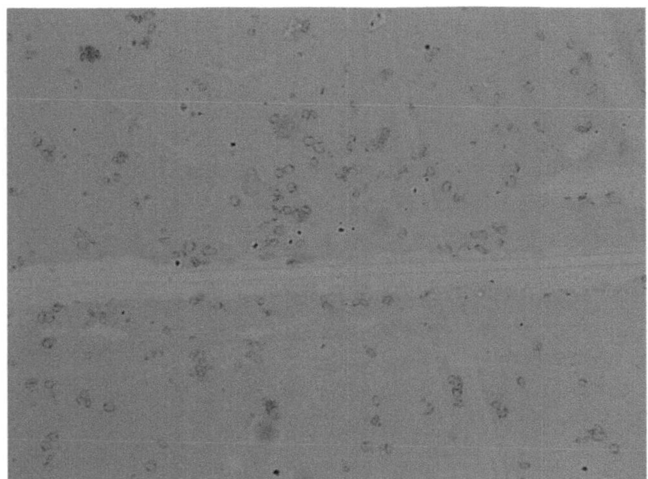

FIGURE 2.11 *Mycoplasma* **(Giemsa stain).** Typical ring-shaped *mycoplasma* cells stain purple.

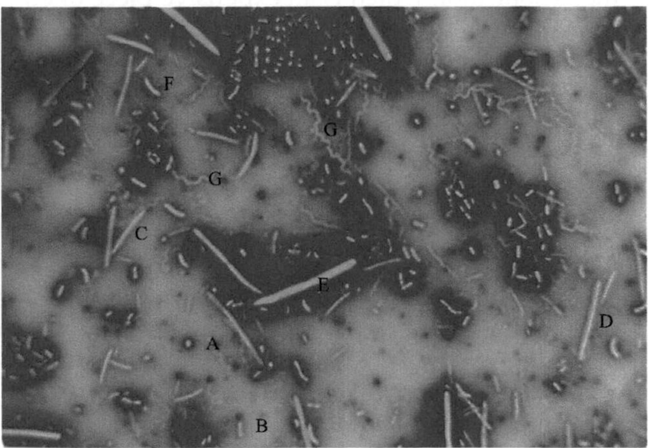

FIGURE 2.12 **Negative Congo red stain of plaque bacteria.** The background is blue while the bacteria remain unstained (bright white with different shapes), creating what is called a negative stain. A. Coccoid cells; B. short bacilli; C. fusiform bacilli; D. long bacilli; E. filamentous bacilli; F. curved rods; G. spirochetes.

Immediately examine the shape and mobility of the protozoans under the microscope. Trophozoites of *T. tenax* are lively pear-shaped parasites that move faster than *Trichomonas vaginalis*. Examination of fresh smears should preferably be performed at room temperatures above 20 °C and completed within 30 min after smear preparation in order to avoid any influence from the lower temperature of the room or the time spent on the slide on the motility of the protozoan.

2.1.8.2 Giemsa Staining

Place a drop of saline solution onto a clean slide, and mix it with the clinical sample (e.g., subgingival plaque). Stain the air-dried specimen with Giemsa stain solution, and examine it with a light microscope under oil immersion.

2.2 ISOLATION, INCUBATION, AND IDENTIFICATION TECHNIQUES

The isolation and identification of oral microorganism can be difficult. Because oral microorganisms are great in number and composed of diverse species, new genera and species are constantly being discovered, while the classification of some previously discovered species changes with time. Therefore, it is currently impossible to isolate and identify all the microbes in an oral specimen. Phenotypic identification of organisms is the most basic and important part of microbiology. Classical methods for bacterial identification require observation of the phenotypic characteristics of a pure bacterial culture, including characteristic colonies (size, color, shape, etc.), cell characteristics (size, shape, arrangement, and stain), special structures (with or without spores, capsule, pili, and flagellum), culture characteristics (sensitivity to oxygen, optimum growth temperature, pH, requirements for nutrients and growth factors, etc.), metabolites, etc. *Bergey's Manual of Systematic Bacteriology* is an authoritative reference book for bacterial isolation.

2.2.1 Collection and Transportation of Samples

2.2.1.1 Sample Collection

Collection of saliva samples: Saliva samples include stimulated salivary and unstimulated salivary samples. Stimulated salivary samples are collected when subjects are chewing paraffin or a rubber block, while unstimulated salivary samples are secreted naturally by the subjects. More saliva is collected by stimulation, but at the same time, it can influence the mucosa, oral plaque, and the oral microflora. Saliva samples for microbiological examination should preferably be fresh unstimulated salivary samples. The best time for collection is early in the morning upon waking and before the teeth are brushed, or alternatively, samples can be collected between meals (around 10 a.m. or 4 p.m.). Subjects should gently rinse their mouths with warm water to remove any food residue. Anywhere from 0.5 to 1 ml of naturally secreted saliva is collected into a sample tube, or saliva samples can be directly taken from the oral cavity using sterile pipette tips.

Collection of plaque samples: As plaque microbes are complex in their composition, plaque samples should be collected in different ways depending on specific clinical requirements and purposes. If necessary, plaque indicators should be used to show dental plaque. Plaque samples can be found on adjacent surfaces and fissures in occlusal surfaces. Before sample collection, subjects should gargle with warm water to remove any food residue. Then, sterile gauze or a yarn ball should be used to absorb saliva while plaque samples are collected. A sterile probe is commonly used in plaque collection from occlusal surface fissures. Plaque found on adjacent surfaces can be collected with sterile probe, dental floss, or fine orthodontic wire. Sterile curettes can be used to collect root surface plaque samples. Plaque samples on the gum or in the gingival margin can be collected using a spoon scaler. Subgingival plaque is divided into attached plaque and unattached plaque. Collection using the MooreOO bacteria taker or using a sterile paper point are currently the most widely used and the easiest way to collect subgingival plaque. Plaque samples from infected root canals are usually collected with sterile paper point.

Collection of other samples of infected tissue: In samples of infected tissue such as lip carbuncle, aerobic bacteria and facultative anaerobes such as *Staphylococcus aureus* are the main pathogens and are generally collected with sterile cotton swabs. Samples of purulent fluid from a periodontal abscess can be collected with sterile syringes. Tissues in the alveolar socket are generally collected as samples of dry socket developed after tooth extraction.

Collection of oral mucosal diseases samples: White membranous materials are usually collected with curettes or cotton swabs. To collect the quantitative samples, filter papers of a specific size are used.

2.2.1.2 *Sample Transportation*

Fungal species, aerobic bacteria, and general facultative anaerobic bacteria can be collected by cotton swab sampling and transported in sterile tubes. For most anaerobic bacteria or microaerophilic bacteria, samples must be sent to the laboratory as soon as possible and maintained in an anaerobic environment. With the exception of pus or saliva samples that can be inserted directly into the sterile rubber stopper of the syringe needle used to collect the sample for transportation, most anaerobic samples must be placed in prereduced culture media and for transportation to the laboratory immediately after acquisition. This is to minimize the death of oxygen-sensitive bacteria during transportation. Chairside inoculation and anaerobic transportation can help improve the detection rate of obligate anaerobic bacteria.

Transportation in prereduced medium requires that samples be placed immediately in a small covered tube containing prereduced liquid transportation medium. In order to reduce the infiltration of oxygen during transportation, sterile liquid paraffin can be placed on the prereduced medium to isolate it from air.

For clinical specimens that cannot be examined in a timely manner or that must be transported over a long distance, anaerobic bags (commercially available) or prereduced liquid medium in a spiral tube sealed with liquid paraffin can be used for transportation.

2.2.2 Suspension and Dilution of Samples

Clinical oral infections are mixed infections involving many different species of bacteria within a concentrated plaque mass. In order to obtain pure cultures of individual bacteria, oral clinical specimens usually require processing and dilution after inoculation.

2.2.2.1 *Sample Suspension in Solution*

Generally, two methods are used for sample suspension: spiral vortex oscillation (Figure 2.13(A)) and ultrasonic dispersion (Figure 2.13(B)). Spiral vortex oscillation is widely used because of its ease of operation and low cost. All that is required is for the sample collection tube to be placed on the vortex generator and allowed to oscillate for 10–20 s. Adding five to six small sterile glass beads (110–150 μm) into the small tube can help improve sample dispersion. Ultrasonic dispersion yields superior sample suspension, but the disadvantage is that a sonicator, which is an expensive piece of equipment, is required, and the microbial detection rate can drop because spirochetes, *Porphyromonas*, *Prevotella*, and other gram-negative bacteria are easily lysed. It is also difficult to avoid oxygen infiltration during ultrasonic dispersion, which can lead to the death of anaerobic bacteria in the sample.

2.2.2.2 *Sample Dilution*

As oral clinical samples are mixed bacterial samples containing a great number and variety of bacteria, proper dilution with the correct diluent must be performed prior to inoculation to obtain single colonies after sample suspension.

The solution in which the sample was transported can be used as a diluent, otherwise, phosphate buffer (pH 7.2) can also be used. A 10-fold dilution series is generally performed. Under aseptic conditions, 0.1 ml of the specimen sample is added to 0.9 ml diluents. After thorough mixing, 0.1 ml of the mixture (10^{-1}) is added to a tube containing 0.9 ml diluent and mixed again. Using this method, 10-fold dilutions are made in series (Figure 2.14). Due to differences in sample bacteria content, different samples require different dilutions. For example, saliva samples should be diluted to 10^{-4}–10^{-6}, gingival groove plaque should be diluted to 10^{-1}–10^{-2}, while mixed plaque should be diluted to 10^{-3}–10^{-5}.

2.2.3 Inoculation and Incubation of Samples

In addition to selecting the appropriate medium for inoculation, other considerations include: degree of dilution before inoculation, method of inoculation, incubation environment, time needed for colony establishment, purpose of incubation, and microbial species.

FIGURE 2.13 (A) Sample dispersion (vortex generator). (B) Sample dispersion (ultrasonic generator).

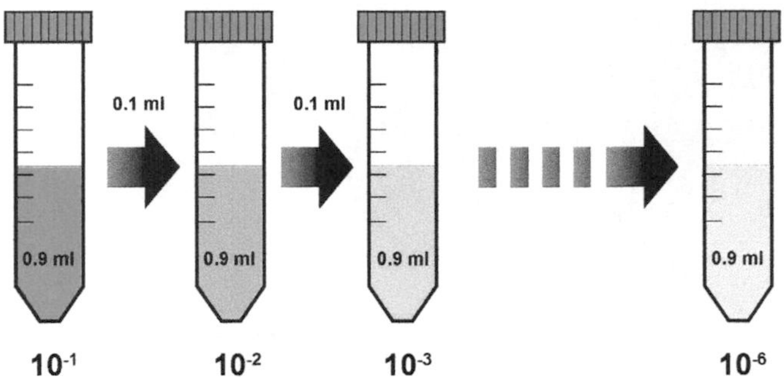

FIGURE 2.14 **Sample dilution (10-fold serial dilution).**

2.2.3.1 *Choice of Medium*

The medium commonly used for oral bacteria include brain-heart infusion (BHI) medium, trypticase soy medium (TSA), and TPY medium. These can be used to cultivate most bacteria from an oral sample. Five percent defibrinated blood (or 5% serum), chlorinated hemoglobin, and vitamin K1 must be supplied to the culture medium for some obligate anaerobic gram-negative bacilli. To culture most aerobic and facultative anaerobic bacteria, the blood agar (BA) is used.

2.2.3.2 *Sample Inoculation*

Spread method, drop method, and spiral inoculation method are used for oral clinical bacteriology samples. Appropriate dilutions of the specimen solution are quantitatively inoculated onto the agar plate.

1. **Spread method:** Ten microliters from appropriately diluted samples are taken with a micropipettor and placed on the surface of the agar medium. The sample is then evenly coated on the agar surface using a sterile glass spreader (triangle rod and L-shaped rod). At the appropriate dilution, each bacterial cell from the specimen should form a single colony after incubation (Figure 2.15(A) and (B)).
2. **Drop method:** Twenty-five or 50 μl samples at the appropriate dilution are dropped onto the surface of the agar using a micropipettor. Then, the plates are placed directly into the dry incubator (Figure 2.15(C) and (D)).
3. **Spiral plater method:** The spiral plater is the most advanced method for inoculating bacteria for the purpose of counting colony-forming units. The sample liquid is automatically diluted and inoculated using a needle tip on the surface of the agar plate by the instrument, and the bacteria colonies grow uniformly along the spiral trajectory after incubation (Figure 2.15(E)). As a result, sample counting and bacterial colony observation are more accurate and reproducible. Details are included in Section 2.4 of this chapter.

2.2.3.3 *Sample Incubation*

Media containing clinical samples are incubated according to the requirements of each specific sample, including oxygen requirement, temperature, and time. Oral clinical specimens such as infected root canal, pericoronitis infection, samples taken after tooth extraction, subgingival periodontitis plaque samples, which are mostly mixed bacterial samples, may involve different oxygen requirements, as there are a variety of microorganisms each with their respective requirements. Some bacteria require incubation under the anaerobic conditions, while others require incubation under aerobic conditions. The variability in culture conditions requires the researcher to become familiar with and to master the growth characteristics of bacteria in the mouth to avoid mistakes in the incubation. Usually, anaerobic cultures grow best in atmospheric conditions with 80% N_2, 10%

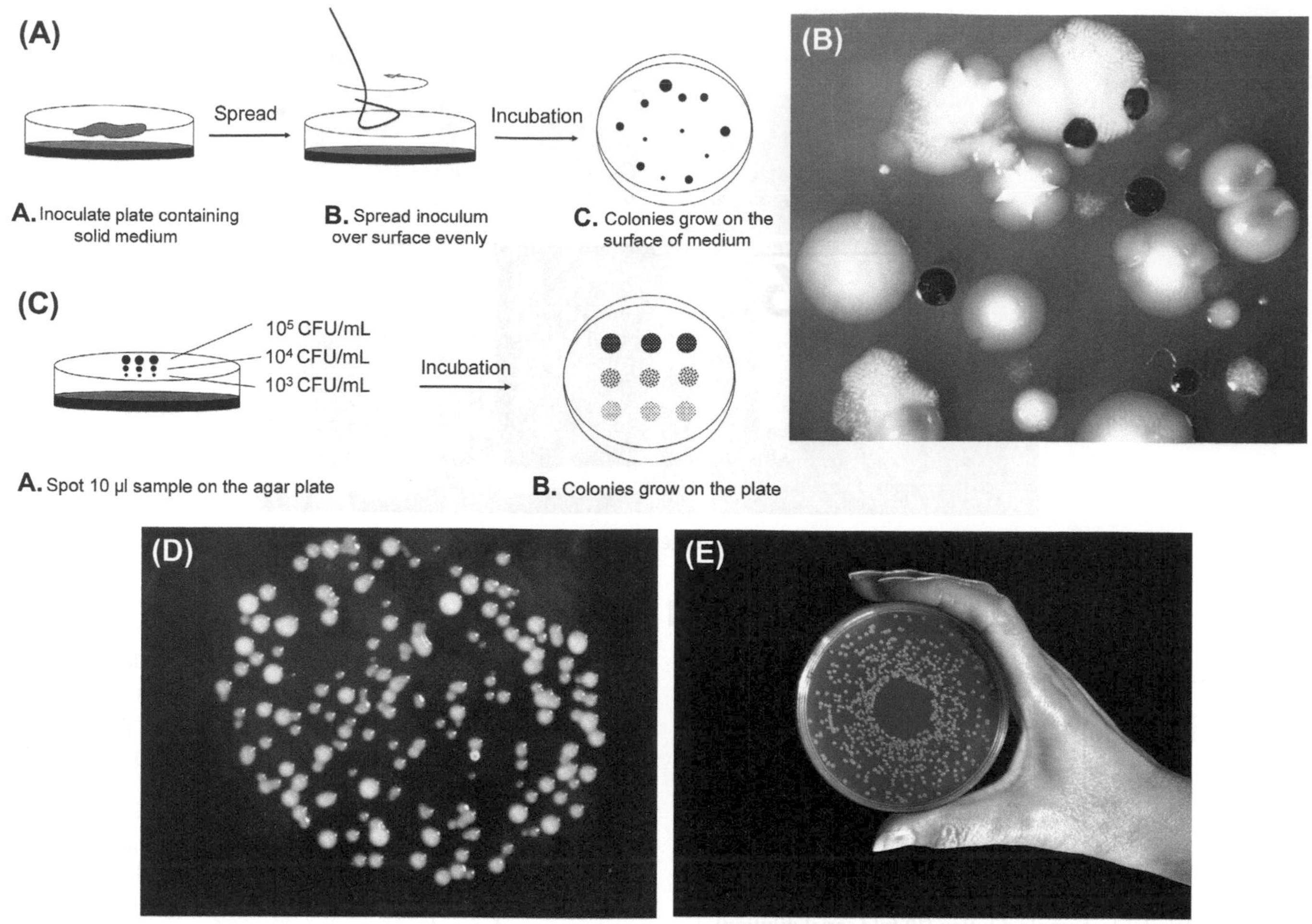

(A)

Spread

Incubation

A. Inoculate plate containing solid medium

B. Spread inoculum over surface evenly

C. Colonies grow on the surface of medium

(B)

(C)

10^5 CFU/mL

10^4 CFU/mL

10^3 CFU/mL

Incubation

A. Spot 10 μl sample on the agar plate

B. Colonies grow on the plate

(D)

(E)

FIGURE 2.15 (A) Spread method protocol. (B) Colonies on a plate using spread method (periodontal bag samples). (C) Drop method protocol. (D) Colonies on a plate using drop method (saliva samples). (E) Colonies on a plate using spiral plater (plaque samples).

CO_2, and 10% H_2, with a temperature of 36–37°C and approximately 48–72h culture time. Some oral bacteria, such as Forsyth Steiner bacterium, *Treponema*, and others, require 1 week of anaerobic culture. Commonly used anaerobic incubation devices include the anaerobic glove box, anaerobic incubator, anaerobic bag, etc. (Figure 2.16).

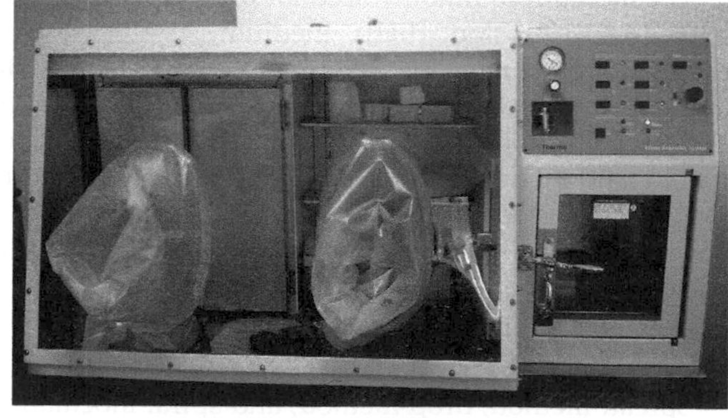

FIGURE 2.16 **Anaerobic glove box.**

The initial steps of bacterial identification involve observing colony morphology following incubation and performing Gram stain. After this, a second round of purification is carried out to further complete phenotypic and genotypic identification using routine protocols for microbial separation and identification. For certain microorganisms, special separation, cultivation, and other identification techniques may be adopted.

2.2.4 Growth Characteristics and Identification

2.2.4.1 *Growth Characteristics*

Bacterial colony morphology can be described in terms of its size, color, shape, growth pattern, and other characteristics. Hemolysis is one of the basic tests used for bacterial identification. Some bacteria produce hemolysis (Figure 2.17(A)–(E)), some produce gas (Figure 2.17(F)), and some exhibit specific growth patterns, such as the migration of *Proteus* (Figure 2.17(G)).

2.2.4.2 *Biochemical Tests*

Biochemical tests are among the most important methods for microbial identification. Routine biochemical tests include tests for carbohydrate fermentation

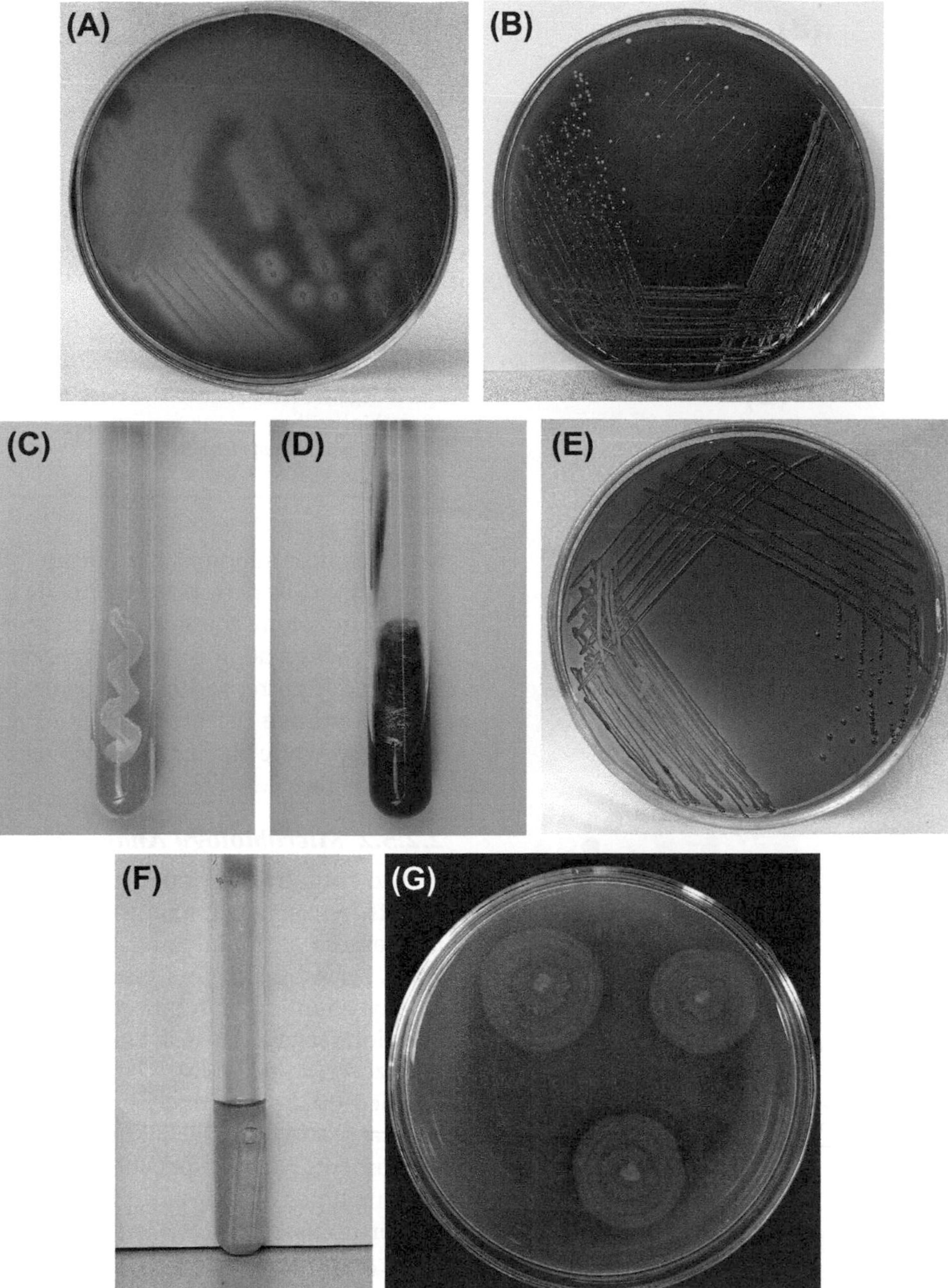

FIGURE 2.17 (A) β-hemolytic reaction. (B) α-hemolytic reaction. (C) White pigment (*Staphylococcus epidermidis*). (D) Green pigment (*Pseudomonas aeruginosa*). (E) Black pigment (*Porphyromonas gingivalis*). (F) Production of gas (*Escherichia coli*). (G) Migrating growth (*Proteus*).

(Figure 2.18(A)), methyl red (Figure 2.18(B)), citric acid utilization (Figure 2.18(C)), and hydrogen sulfide production (Figure 2.18(D)).

Microbial biochemistry tests shorten the time required to identify microbes, reduce costs, and ensure or enhance the accuracy of identification of an unknown sample. It is the fastest developing trend in microbial identification. In recent years, the rapid commercial test kits for anaerobic bacteria have become available.

The most representative biochemical test kits are the Minitek identification system using paper substrates, API-20A system using dry powder substrates, PIZYMAN-IDENT rapid enzyme activity assay system using primary materials, RaPID-ANA systems, and fully automated microbial identification systems.

The aforementioned microbial biochemistry reaction plate includes 30 biochemical matrices and their related biochemical test indicators, phosphate buffered saline (PBS), bacterial turbidity standard tube, and eight identification series (Table 2.1).

Due to the different types of experiments performed, the readouts for results are different. For example, esculin

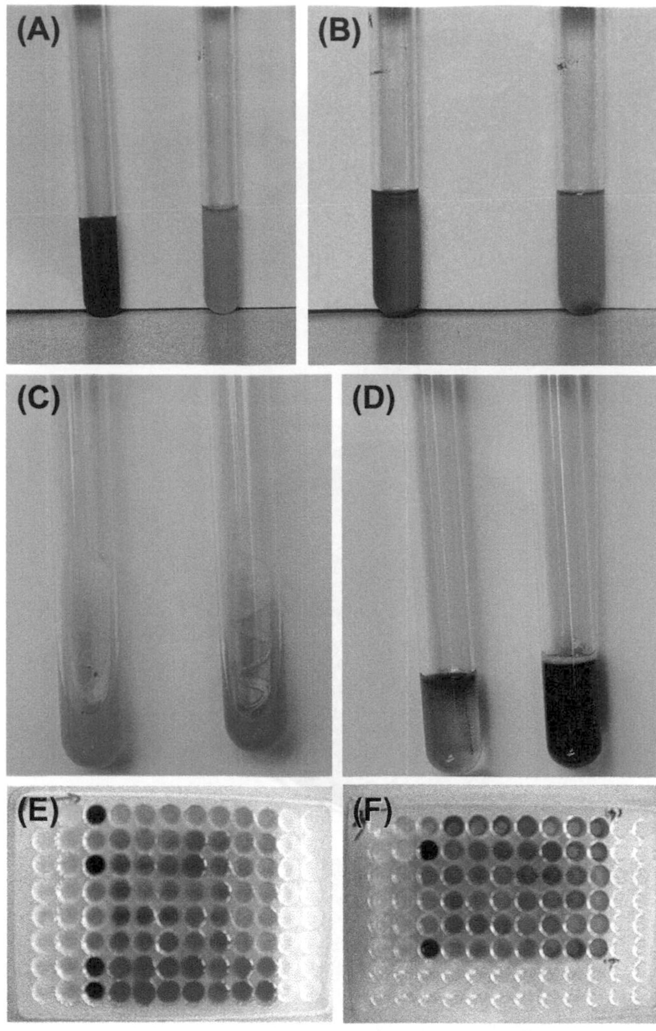

FIGURE 2.18 (A) Carbohydrate fermentation test. (B) Methyl red test. (C) Citric acid utilization test. (D) Hydrogen sulfide production test. (E) Gram-positive anaerobic cocci series II (*Streptococcus* series) (*Streptococcus mutans*). (F) Gram-negative anaerobic nonspore bacillus series II (black pigment produced) (*Porphyromonas gingivalis*).

hydrolysis can be directly observed: black is positive, while colorless is negative. For sugar and alcohol fermentation acid test, BM (bromothymol blue-methyl red, BTB-MR) must be added to the result as a pH indicator. Red or yellow indicates a positive reaction, while green or blue indicates a negative reaction (Figure 2.18(E) and (F)).

2.2.5 Instruments for Microbiological Identification

2.2.5.1 Spiral Plater

A spiral plater (Figure 2.19(A)) is used to inoculate plates to determine viable bacteria count. The working principle behind a spiral plater is the use of a tip to dispense the liquid inoculum onto a Petri dish in a spiral pattern. The spiral plater deposits a known volume of sample onto a rotating agar plate so that the sample forms a spiral pattern with highest concentration at the center and lowest concentration at the outside of the spiral. Colony counting can be performed manually counting or by using automatic colony counters.

With a high degree of automation, simplicity of operation, and high reproducibility, spiral platers can significantly improve the efficiency and accuracy of bacteria counting while saving manpower, time, and culture space (Figure 2.19(B)).

TABLE 2.1 Microbial Biochemistry Test Identification Series

Main classification series	Sub-classification series
Gram-positive anaerobic cocci	I. Staphylococci and micrococci II. *Streptococcus*
Gram-negative anaerobic cocci	
Gram-positive anaerobic nonspore bacillus	
Gram-negative anaerobic nonspore bacillus	I. Does not produce black pigment II. Produces black pigment
Gram-negative anaerobic *Clostridium* or *Enterobacter*	
Gram-negative facultative anaerobic bacillus	
Gram-negative *Campylobacter*	
Clostridium	

2.2.5.2 Microbiology Analyzer

The automated microbiology analyzer provides results by capturing images with a high-definition digital camera and analyzing them using its built-in software. Microbiology analyzers can be used for automatic colony counting, inhibition zone measurement to test antibiotic sensitivity test, and measurement of the hemolytic zone (Figure 2.20).

Microbiology analyzers are simple to operate, fast, accurate, and provide highly reproducible results. The instruments have a special lighting system suitable for various types of agar and plates. Other features include powerful analysis software and vivid full-color images that can be saved digitally or printed out, with results labeled beside the images and automatically recorded.

2.2.5.3 Microbial Identification System

Microbial cells produce different enzymes during their metabolism of different carbon sources. The MicroStation automated microbial identification system is based on the differences in color and in turbidity that occur when these enzymes react with four azole substances (e.g., TTC, TV, etc.). With the use of a unique technology that detects the characteristic fingerprint of each microorganism and based on a large number of experiments and mathematical models, the corresponding database between the fingerprints and microbial species has been established. Identification results can be derived through comparisons between the unidentified microbial species and the reference database by software (Figure 2.21).

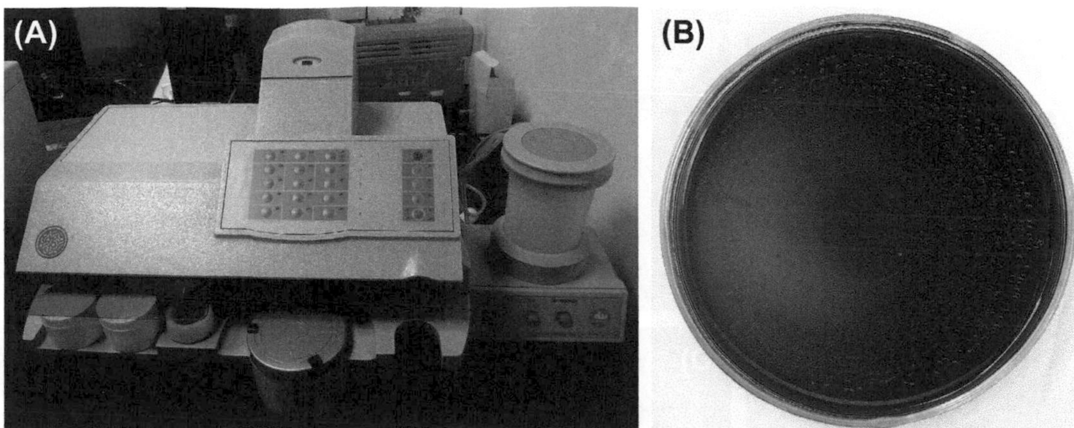

FIGURE 2.19 (A) WASP spiral plater. (B) The colonies on a plate prepared by spiral plater. After inoculation on a 9-cm plate, the sample will be 1000-fold diluted on the outside track and the colonies will be evenly distributed.

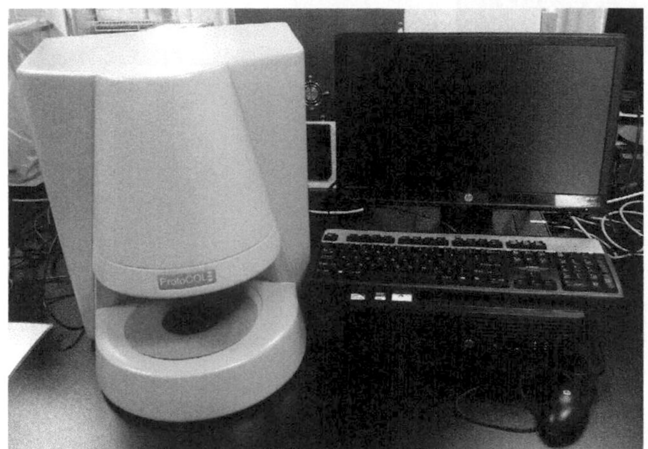

FIGURE 2.20 **Synbiosis automated microbiology analyzer.**

FIGURE 2.21 **MicroStation automated microbial identification system.**

The MicroStation automated microbial identification system is used for microbial identification in clinical settings, food, dairy, pharmaceutical, and cosmetics industries, and environmental microbial identification (in rivers, oceans, plants, animals, and insects). It can also be used to analyze microbial communities and aid in ecological research in the analysis of carbon utilization and metabolism. There are four types of microplates specifically designed for the analysis of microbial communities and ecosystems research.

2.3 MICROSCOPY TECHNIQUES

2.3.1 Stereomicroscopy

The stereomicroscope (Figure 2.22(A)) is an optical microscope that produces a three-dimensional visualization of the sample being examined. The instrument is also known as a stereoscopic microscope or dissecting microscope.

The optical structure of the stereomicroscope includes one shared primary objective and two sets of intermediate objective lenses or zoom lenses. Two beams are separated by intermediate objectives at an angle of about 12–15°, called the stereo angle, and then are imaged by their corresponding optical lens. The beams are not parallel but at an angle, providing a three-dimensional image to both eyes by two separate optical paths. The magnification changes when the distance between intermediate objectives is changed.

The digital imaging system connected to the computer to analyze and process images is composed of the stereomicroscope, a variety of digital ports, digital cameras, electronic optical lenses, and the image analysis software.

Before using the stereomicroscope, the instrument should be adjusted for focus, diopter, and inter-pupil distance for each user in order to acquire the best image.

The stereomicroscope is used to observe microbial colony morphology (Figure 2.22(B)–(F)) and microbial distribution (Figure 2.22(G)–(J)).

2.3.2 Scanning Electron Microscopy

The scanning electron microscope (SEM, Figure 2.23) is mainly used to observe the topography of

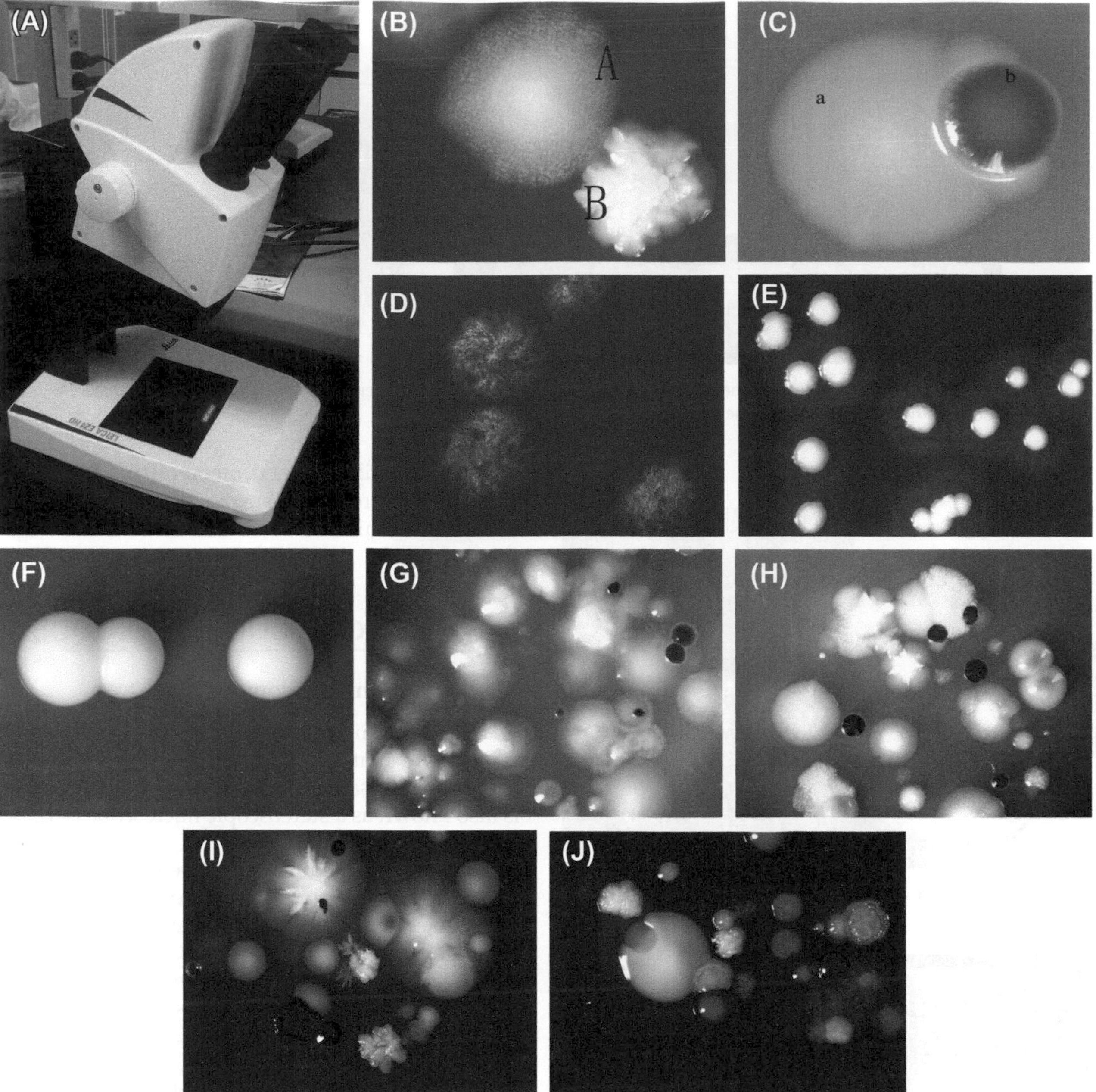

FIGURE 2.22 (A) Stereomicroscope. (B) Isolated bacteria from dental plaque sample a: *Actinomyces israelii* colonies; b: *Fusobacterium nucleatum* colonies (BHI blood agar, anaerobic culture for 48h, stereomicroscopy). (C) Characteristic colonies of saliva bacteria a: *Prevotella oris* (pink colonies); b: *Actinomyces odontolyticus* (red colonies) (BHI blood agar, anaerobic culture for 48h, stereomicroscopy). (D) Diffusively growing colonies of *Capnocytophaga sputigena* (BHI blood agar, stereomicroscope). (E) α-hemolytic colonies of *Streptococcus gordonii* (BHI blood agar, stereomicroscope). (F) *Candida albicans* colonies (BHI blood agar, stereomicroscope). (G) Bacterial colonies from saliva sample (BHI blood agar, stereomicroscope). (H) Bacterial colonies from gingival margin sample (BHI blood agar, stereomicroscope). (I) Bacterial colonies from nonadhesive subgingival plaque (BHI blood agar, stereomicroscope). (J) Bacterial colonies from periodontal pocket sample (BHI blood agar, stereomicroscope).

the cells in the samples over a large range of magnification. Sample preparation for SEM is simple. It is adaptable to various samples and does not require producing ultra-thin slices. SEM is already a routine method in medical research and is especially crucial for studies on the morphologics and interactions of oral bacteria.

SEM can be used to analyze and interpret observations on a micron or nanometer scale. The resolution of a field emission scanning electron microscope can reach as low as 1 nm. Another important feature of the scanning electron microscope is that it can be used to observe and analyze samples three-dimensionally due to its deep depth of field. The greater the depth of field,

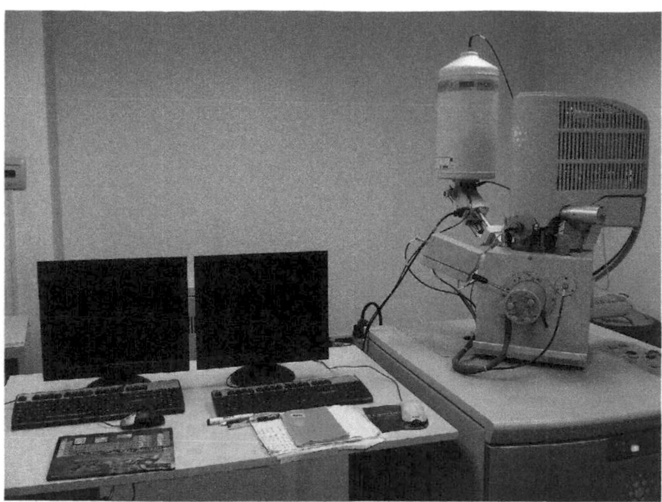

FIGURE 2.23 **Scanning electron microscope.**

the more sample information is provided. In microbial identification, SEM is utilized to observe and detect surface morphology and structural characteristics of microbial cells.

2.3.2.1 *Mechanism*

The scanning electron microscope is used to scan sample areas or microvolumes with a fine focused beam of electrons, producing various signals including secondary electrons, back-scattered electrons, Auger electrons, characteristic X-rays, and photons carrying different levels of energy. When the electron beam scans the sample surface, the signals will change according to the surface topography. The limited emission of secondary electrons within the volume close to the electron focusing area results in high image resolution. The three-dimensional appearance of images comes from the deep depth of field and shadow effect of secondary electron contrast.

2.3.2.2 *Operating Procedure*

1. Bacterial sample preparation

Culture bacteria at 37 °C for 48 h, identify cells as pure culture by morphological and biochemical tests, make bacterial suspension with 0.2 mol/l phosphate buffer (pH 7.2).

2. Sample washing and fixation

Wash sample twice with phosphate buffer or saline, fix sample for 2 h (at 4 °C) with 2.5% or 3% glutaraldehyde, then wash twice with phosphate buffer or saline.

3. Gradual dehydration

Dehydrate samples with ethanol using a concentration gradient: 30% ethanol for 20 min, 50% ethanol for 20 min, 70% ethanol for 20 min, 80% ethanol for 20 min, 90% ethanol for 20 min, and finally 100% ethanol for 20 min twice.

4. Liquid exchange

Place dehydrated samples into 100% amyl acetate solution for 20 min for exchange.

5. Critical point drying

Place samples into a CO_2 critical point dryer for CO_2 critical point drying.

6. Metal coating

Coat the dried samples with ion sputter coater.

2.3.2.3 *Sample Observation*

Observe processed samples using the scanning electron microscope. The authors used the Inspect F field emission scanning electron microscope to observe surface morphologies and structural characteristics of oral microbial cells, as shown in the following examples (Figure 2.24).

2.3.3 Transmission Electron Microscopy

Transmission electron microscopy (TEM) (Figure 2.25) is mainly used to observe the cell's internal structures using ultrathin sections.

2.3.3.1 *Preparation of Bacterial Samples*

Oral bacteria culture is centrifuged at 3000 r/min for 20 min, and the supernatant is removed. The pellet is washed three times with saline, and a small amount of serum is added. The bacteria are harvested by centrifugation at 3000 r/min for 20 min, and the supernatant is removed. The pellet is rinsed with 2.5% glutaraldehyde (prepared with sodium cacodylate buffer), sodium cacodylate buffer, and fixed with 1% osmic acid. The sample was sealed with a series of 50%, 70%, 90%, and 100% ethanol dehydration and embedded with epoxy resin to make ultrathin slices.

2.3.3.2 *Section*

Ultrathin sections are defined as sections with a thickness between 10 and 100 nm. The technology with which these slices are made is called ultrathin slice technology and includes the collecting samples, fixing, rinsing, dehydrating, penetrating, embedding, sectioning, and dyeing. Compared with optical microscopy, the process of sample preparation for TEM is more sophisticated and stringent.

2.3.3.3 *Uranium Staining and Observation*

Professional workers use a TEM to observe the inner structure of cells via ultrathin sections (Figure 2.26).

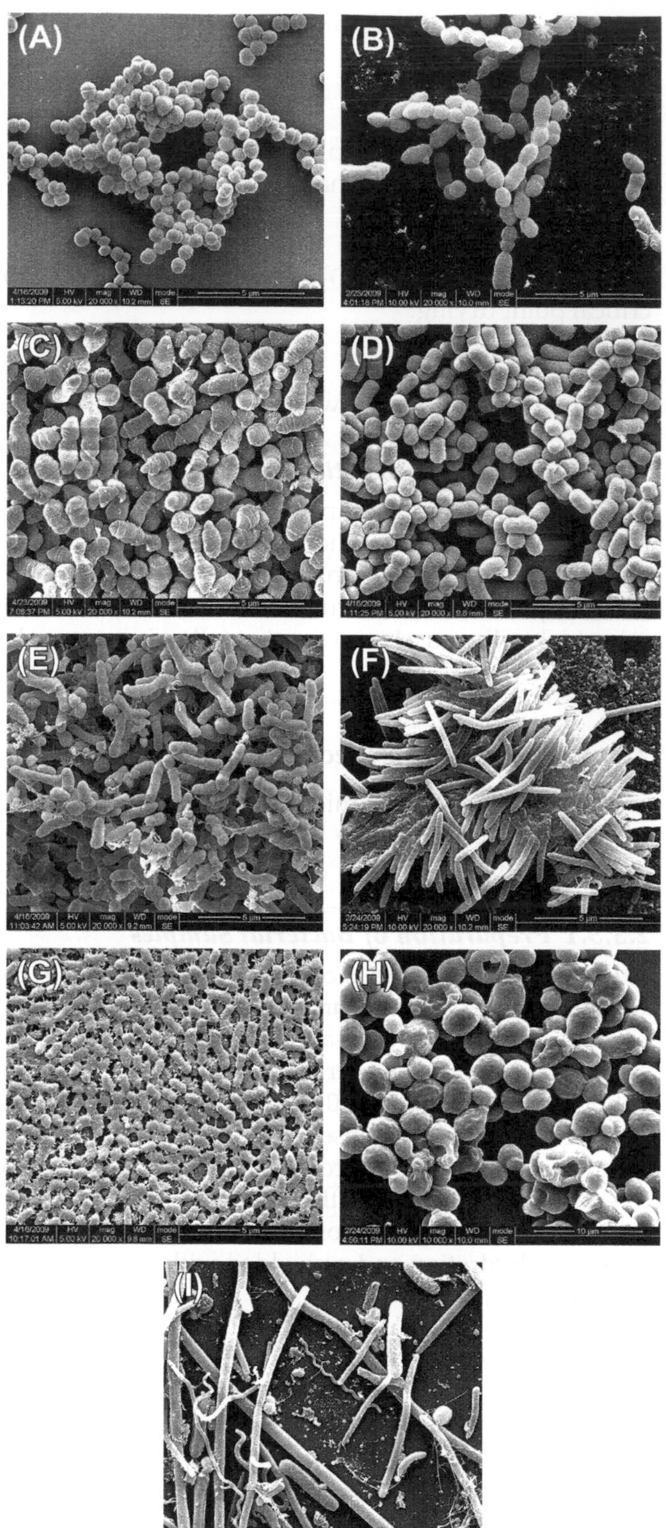

FIGURE 2.24 (A) Proliferating cells of β-hemolytic *Streptococcus* (SEM). (B) Division phase of α-hemolytic *Streptococcus* cells (SEM). (C) Proliferating cells of *Streptococcus gordonii* (SEM). (D) Proliferating state of *Lactobacillus fermentum* cell (SEM). (E) Self-aggregating cells of *Rothia dentocariosa* (SEM). (F) Cell aggregation of *Capnocytophaga sputigena* (SEM). (G) Massive extracellular matrix of *Porphyromonas gingivalis* (SEM). (H) Budding cells of *Candida albicans* (SEM). (I) Spirochetes in gingival margin plaque (SEM).

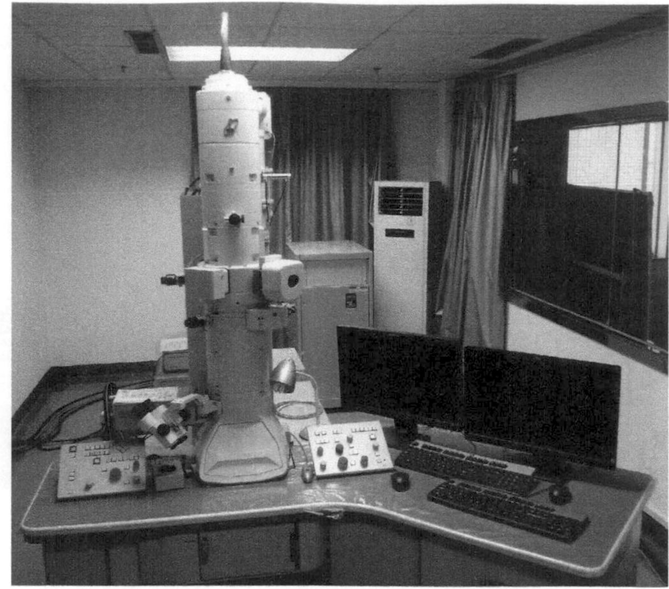

FIGURE 2.25 **Transmission electron microscope.**

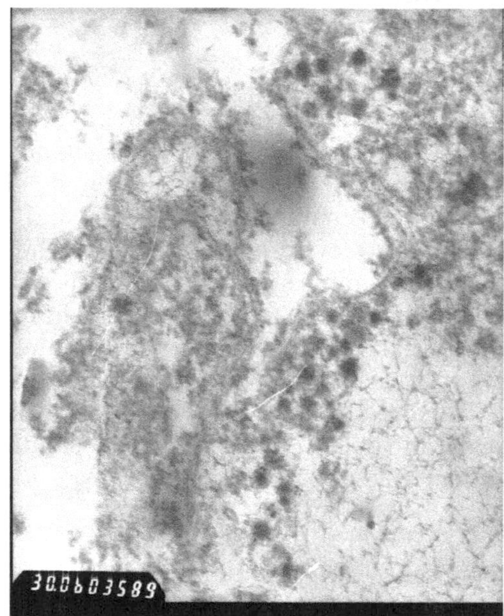

FIGURE 2.26 **Adenovirus Ad-hTR-si in HEK293 cell (TEM).**

2.3.4 Confocal Laser Scanning Microscopy

Confocal laser scanning microscopy (CLSM) was developed in the late 1980s. With the unparalleled advantages of high resolution, ease of sample preparation, dynamic recording without damaging the living cell, acquisition of three-dimensional sample images though tomography and 3D reconstruction, and spatial positioning of the target, CLSM has become widely used in almost all areas of cell research in medicine and biology (Figure 2.27).

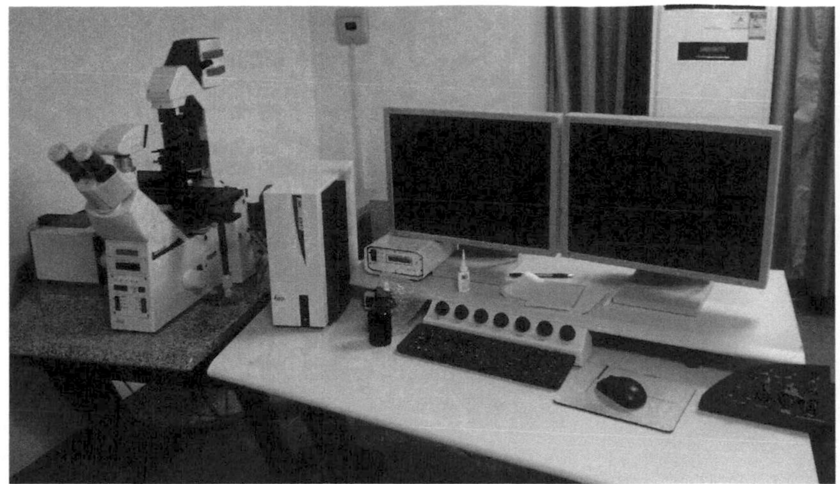

FIGURE 2.27 **Confocal laser scanning microscopy.**

2.3.4.1 *Principles of CLSM*

A confocal laser scanning microscope is made up of the optical system, the laser light source, the detection system, and the scanning device.

The optics behind this type of imaging is a laser emitted from the light source that becomes a parallel beam of expanded diameter when it passes through the pinhole aperture, encounters the dichromatic mirror, and is reflected onto the objective lens. The light beam is reflected 90° when it hits the dichromatic mirror and is focused onto the desired focal plane on the sample when it passes through the objective lens. The fluorescence-emitting sample fluoresces in all directions under excitation from the laser. Part of the fluorescence becomes focused at the focal point of the objective once it passes through the objective lens, dichromatic mirror, and focusing lens. The fluorescent light passes through a pinhole at the focal point and can then be picked up by the detector. When a laser scans the sample point by point, the photomultiplier tube behind the pinhole receives the corresponding point-by-point confocal optical image. Accordingly, different focal planes within the sample and optical cross-sectional images (also known as optical sectioning) can be analyzed one by one. Using computer image processing and three-dimensional image reconstruction software, a high-resolution three-dimensional image can be obtained from the sample. Cell structure, cell content, and dynamic changes can be analyzed by continuous scanning on the same plane. The optical path of a confocal laser scanning microscope is shown in Figure 2.28.

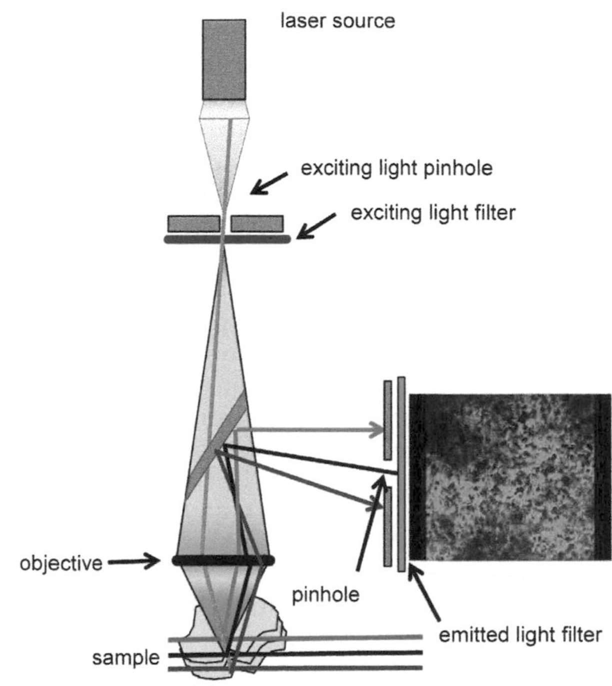

FIGURE 2.28 **The optical path of a confocal light scanning microscope.**

2.3.4.2 *Application in Dental Plaque Research*

Through a special fluorescent staining, dental plaque in its natural hydration status can be studied directly. Through this process, dead and viable bacteria in dental plaque can be observed in situ, and the relationship between bacteria during the formation of dental plaque can be observed as well. Images of a single cell, a group of cells, or different levels within tissues can be obtained by scanning biofilm of a given thickness continuously using CLSM. A complete 3D structure of plaque can be generated by 3D image reconstruction (Figure 2.29).

Compared with SEM, CLSM requires fewer sample preparation steps before the composition and structure of dental plaque can be studied. During sample preparation in SEM, structural damage to the sample can accumulate during steps such as dehydration, fixing, embedding, and dyeing.

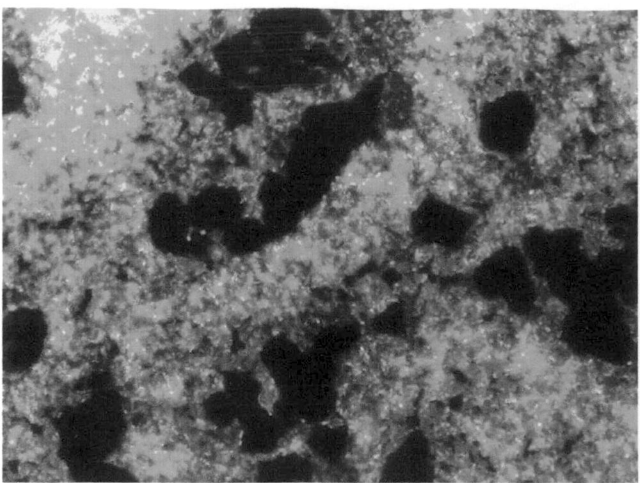

FIGURE 2.29 **Biofilm of *Streptococcus mutans* (CLSM).**

Moreover, CLSM can be used to observe structures, specific molecules, and biological ionic changes in living cells. The technique can also be used to track structural changes and physiological processes within living cells over time to detect changes under the natural state or after stimulation by certain factors. Quantitative and qualitative measurements can be made regarding the perimeter or area of a sample, average fluorescence intensity of cells, in situ determination of cellular contents, composition and distribution of lysosomes, mitochondria, endoplasmic reticulum, cytoskeleton, structural proteins, DNA, RNA, enzymes, cellular receptors, etc. Physiological signals can be dynamically monitored, including quantitative analysis of various ions (mainly calcium ions) with millisecond time resolutions using fluorescent probes.

2.4 ORAL MICROECOLOGY TECHNIQUES

2.4.1 Methods for Measuring Microbial Growth Curves

In microbiology, growth generally refers to the increase in cell number, rather than the volume, of a single cell. The rate of growth, defined as the change in cell number over time, is closely related to the growth cycle, which can be divided into four phases: lag phase, logarithmic (or exponential) phase, stationary phase, and decline phase, according to the characteristics of the growth curve. The rate of growth differs during these distinct phases. We commonly use the growth curve to portray dynamic changes in bacterial cell number during the growth cycle (Figure 2.30).

Lag phase: Bacterial reproduction is slow.

Logarithmic phase: Bacterial reproduction is fast, and the number of viable cells exponentially increases.

Stationary phase: The rate of growth decreases gradually, while the number of dead cells increases.

Decline phase: The growth gradually slows down until it stops completely, the dead cells outnumber viable ones, and the cells show abnormal phenotypes and autolysis.

2.4.1.1 Protocol

Inoculate cells into fresh medium and cultivate under desired growth conditions. The number of bacterial cells will constantly change during the growth cycle. Graph the growth curve using the number of bacterial cells as the Y-axis and the time of growth as the X-axis.

2.4.1.2 Methods

Turbidimetry and viable count methods are commonly used to determine the growth curve.

1. **Turbidimetry:** After inoculation, measure the optical density (OD) of the cell culture during cultivation. Graph the growth curve using the OD value as the Y-axis and the cultivation time as the X-axis.
2. **Viable count method:** In microbial ecology research, the number of viable cells reflects dynamic changes in bacterial growth. Generally, the cells are plated and colonies are counted (more details in "methods for measuring colony forming units" below) to measure the number of viable cells, which are the only cells in the culture that can undergo cell division and reproduce. After inoculation and a certain period of cultivation, inoculate a certain volume of the cell culture onto the agar plate. Graph the growth curve using the logarithm of the number of colonies as the Y-axis and the growth time as the X-axis.

2.4.2 Methods for Measuring Colony Forming Units

Quantification of spatiotemporal changes in the number of organisms in an ecosystem, especially the percentage of viable cells, is an important part of microecological studies. Plate colony-counting methods have been adopted for widespread use internationally to measure the percentage of viable cells in a sample, rather than more traditional absolute viable count methods. Most commonly used methods for plating samples include spread method, drop method, and spiral plating. In addition, there are several reports of the pouring method being a viable way to measure colony forming units.

2.4.2.1 Protocol

After collecting samples from the desired locations and at the appropriate time points, the sample is transported to the laboratory under suitable conditions, suspended, and diluted to an appropriate concentration. The diluted cells should then be quantitatively inoculated onto a

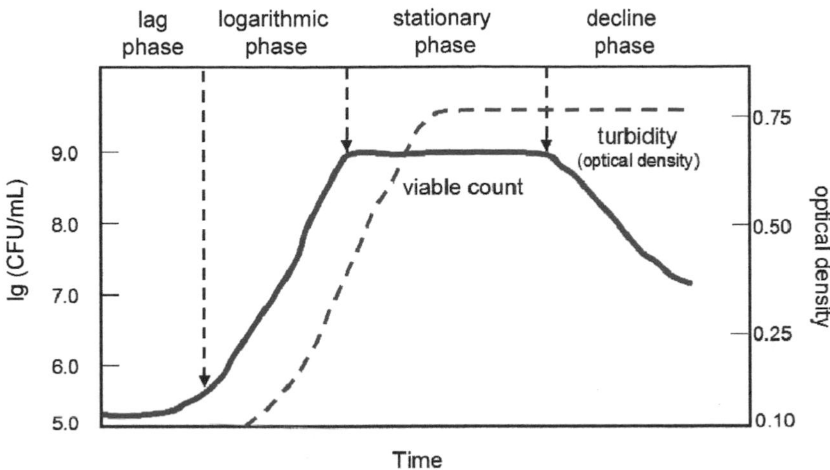

FIGURE 2.30 **Typical example of a growth curve.**

suitable agar plate and placed in a 37 °C incubator after the correct atmospheric conditions and culture time are determined according to the species. After incubation, every viable cell in the sample will form a visible colony. Count the number of colonies and calculate the number of viable cells in the sample based on the degree of dilution.

2.4.2.2 Methods

Plating methods for colony counting can be divided into the spread method, drop method, pour method, and the spiral plater method, based on the approach taken to inoculate a liquid culture of bacteria onto the plate. They share the same basic protocols for sample collection, transportation, dilution, inoculation, incubation, and colony counting, but are different when it comes to the specifics of sample dilution and inoculation. For more details on the spread method, drop method, and spiral plater method, see Section 2.2.3.2. These are the most commonly used plate-based colony-counting methods. In addition, several published reports have shown that the pour method may be a viable option as well (Figure 2.31(A)). Figure 2.31(B)–(E) show the appearance of colonies formed by the spread method and drop method.

The spiral plater uses a spiral plater instrument to automatically perform sample dilution and inoculation. After incubation the colonies grow along the spiral trajectory and colony counting can be performed manually or using an automatic colony counter. For more details, see Section 2.2.3.2.

1. **Pouring method:** Mix the bacterial diluent and the 50 °C soft agar in a sterile bottle. Pour the mixture into a sterile plate (90 mm in diameter), making sure that the mixture is evenly distributed. After incubation, count the colonies on the surface and within the agar medium.
2. **Drop method:** Drop a certain volume of the appropriately diluted sample (10^5 CFU/ml, 10^4 CFU/ml, 10^3 CFU/ml) onto the surface of agar plates (three replicates for each concentration). Count the colonies after incubation.

2.4.3 Measurement of Adhesion Strength and Rate of Adhesion Inhibition

Microbial adhesion is the basis of colonization and pathogenesis. Measurement of adhesion strength and the rate of adhesion inhibition contribute to the understanding of mechanisms of bacterial pathogenesis and control.

2.4.3.1 Measurement of Adhesion Strength

Medium adhesion: Choose slide, glass rod, hydroxylapatite, or teeth as medium for adhesion.

Method: Collect adhesive substance on the surface of the medium surface to evaluate adhesion by measuring the quantity of bacteria (Figure 2.32). By adding artificial saliva or collagen solution into the culture, or by coating the medium for adhesion with saliva or collagen, adhesion strength can be improved and the quality of the results can be improved.

2.4.3.2 Measuring the Rate of Adhesion Inhibition

Measuring the rate of adhesion inhibition allows the researcher to quantify the inhibition of bacterial adhesion by drugs or other reagents. The method is identical to that used to measure adhesion described in the previous section. However, inhibitors of adhesion must be added to the experimental group (Figure 2.33).

2.4.4 Techniques for the Detection of Plaque Biofilm

In nature, the vast majority of bacteria are attached to the surface of living and inanimate objects, where they survive and grow in the form of biofilm. Biofilms are

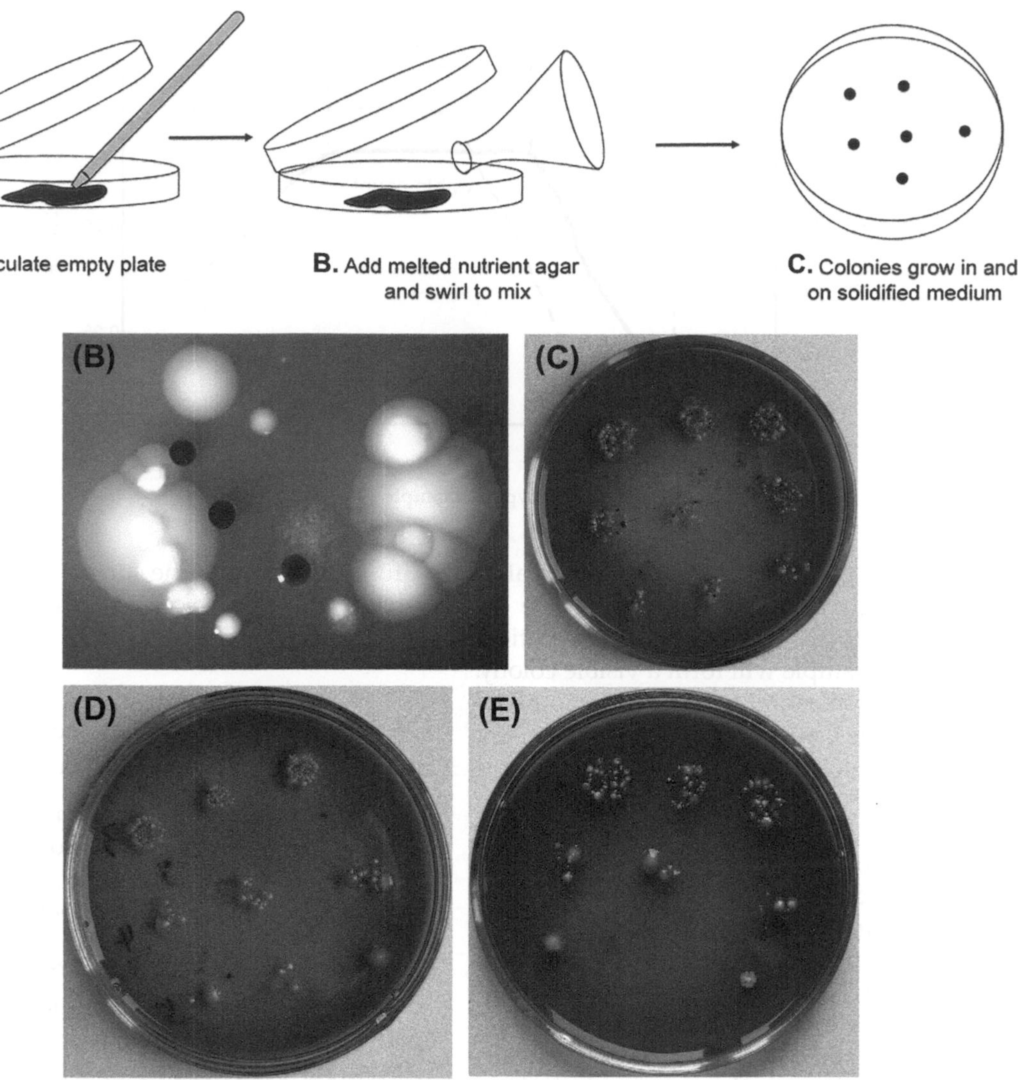

FIGURE 2.31 (A) Protocol of the pouring method. (B) Colonies on a plate prepared with the spread method (gingival marginal plaque sample). (C) Colonies on a plate prepared with the drop method (plaque sample). (D) Colonies on a plate prepared with the drop method (saliva sample). (E) Colonies on a plate prepared with the drop method (subgingival plaque sample).

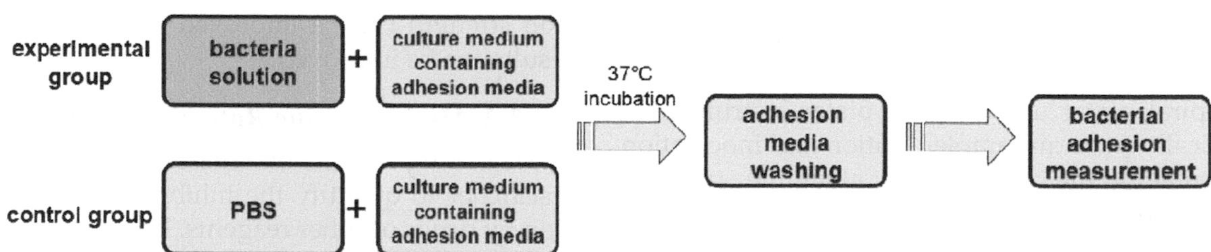

FIGURE 2.32 **Protocol for measuring adhesion strength.** Place medium into a known quantity of culture and add a known quantity of bacterial suspension. After incubation under appropriate conditions, wash the adhesion medium three or four times with KCl buffer to remove nonadhered surface bacteria. Measure adhesion capacity using an isotope liquid scintillation detector, and express the adhesion capacity in scintillation counts per minute (CPM). The adhesion rate (%) = (experimental group CPM – negative control CPM/positive control CPM – negative control CPM) × 100%.

groups of bacteria attached to a surface and enclosed in a secreted adhesive matrix and are functional, interacting, and growing bacterial communities. Dental plaque is a typical kind of biofilm formed on the tooth surface by oral microbes. Microbes inside the biofilm survive as a group with interdependence and mutual competition. They also form a complex ecological relationship. Technologies used to detect biofilm are used to analyze the

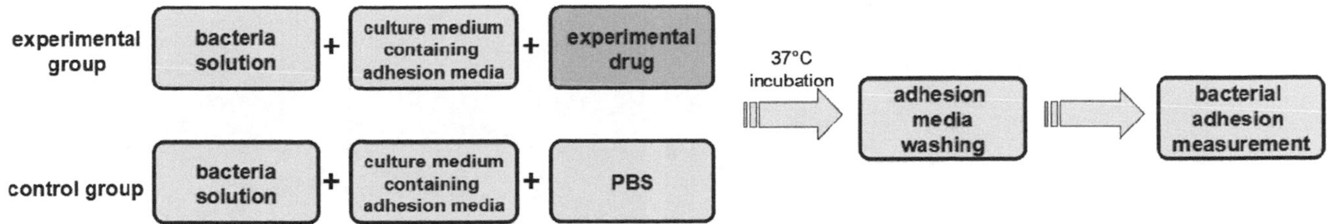

FIGURE 2.33 **Protocol for measuring the rate of adhesion inhibition.** Adhesion inhibition rate = (1−experimental CPM − negative control CPM/positive control CPM − negative control CPM) × 100%. When the CPM of the experimental group is less than that of the negative control group, the rate of adhesion inhibition is defined to be 100%.

natural state of bacteria, the relationship between different bacterial species and the host, pathogenic mechanisms, effects of antibacterial reagents, etc.

2.4.4.1 *Biofilm Formation Assay*

The biofilm formation assay is the basis of a series of biofilm detection technologies and can be used to detect single species biofilm or mixed species biofilm formation as well as their structural characteristics, conditions for formation, and factors that influence their formation. The assay lays the foundation for further studies on relationships between bacterial species, mechanisms of pathogenesis, and preventative measures. The early steps of the assay involve steps that are identical to those in adhesion strength measurements, particularly when it comes to bacterial culture. However, the structure and other features of biofilms are evaluated using SEM and confocal laser scanning microscope.

Currently, there is a commercially available microwell plate for biofilm detection named the 96 MBECTM-Device (MBEC Biofilm Technology Ltd., Calgary, Alberta, Canada, U.S. Patent), shown in Figure 2.34.

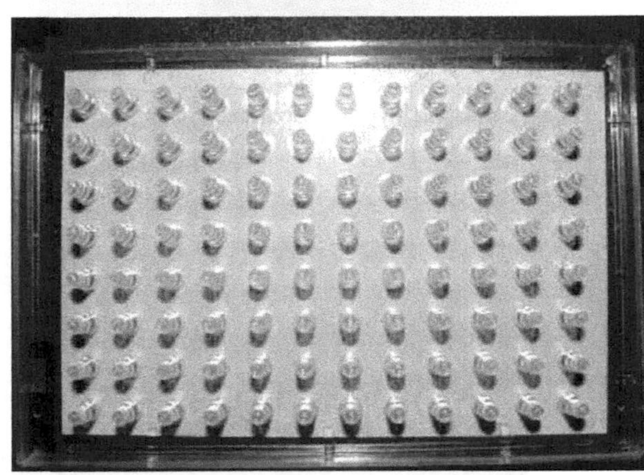

FIGURE 2.34 **Biofilm detection micro-well plate (MBEC™-Device).** The MBEC™-Device has 96 removable pegs (adhesion medium) in its plate cover that sit in the 96 micro wells in the corresponding base. Biofilm formation on the 96 pegs can be detected at the same time, but every peg can also be removed when the biofilm on that peg must be analyzed independently.

2.4.4.2 *Biofilm Detection and Analysis*

By SEM and laser confocal microscopy, characteristic biofilm morphology, structure, and extracellular polymers can be detected. Using SEM, biofilm growth can be observed at different times and under different conditions (Figure 2.35(A)–(C)).

The combined use of fluorescence staining and laser confocal microscopy is a common method to study biofilm structure and extracellular polysaccharides. Biofilm structure can be observed clearly with this technology (Figure 2.35(D)).

2.5 ORAL MICROBIOME TECHNIQUES

The oral microbiome refers specifically to microorganisms (e.g., bacteria, archaea, fungi, *Mycoplasma*, protozoa, and viruses) that inhabit the human oral cavity. Among them, oral bacteria make up the largest proportion of the oral microbiome and are also the most complex in organization. So far, more than 250 oral bacteria species have been isolated, cultivated, and named. Over 450 species have been identified by culture-independent approaches. These bacteria can be classified into different categories based their Gram stain results (gram-positive or gram-negative bacteria), their shape (coccus, bacillus, or spirochetes), and their tolerance to oxygen (aerobic, facultative anaerobes, microaerobic, or obligate anaerobes).

2.5.1 Denaturing Gradient Gel Electrophoresis and Temperature Gradient Gel Electrophoresis

Denaturing gradient gel electrophoresis (DGGE) and temperature gradient gel electrophoresis (TGGE) are forms of gel electrophoresis that use either a chemical gradient or a temperature gradient to separate samples as they move across an acrylamide gel. DGGE

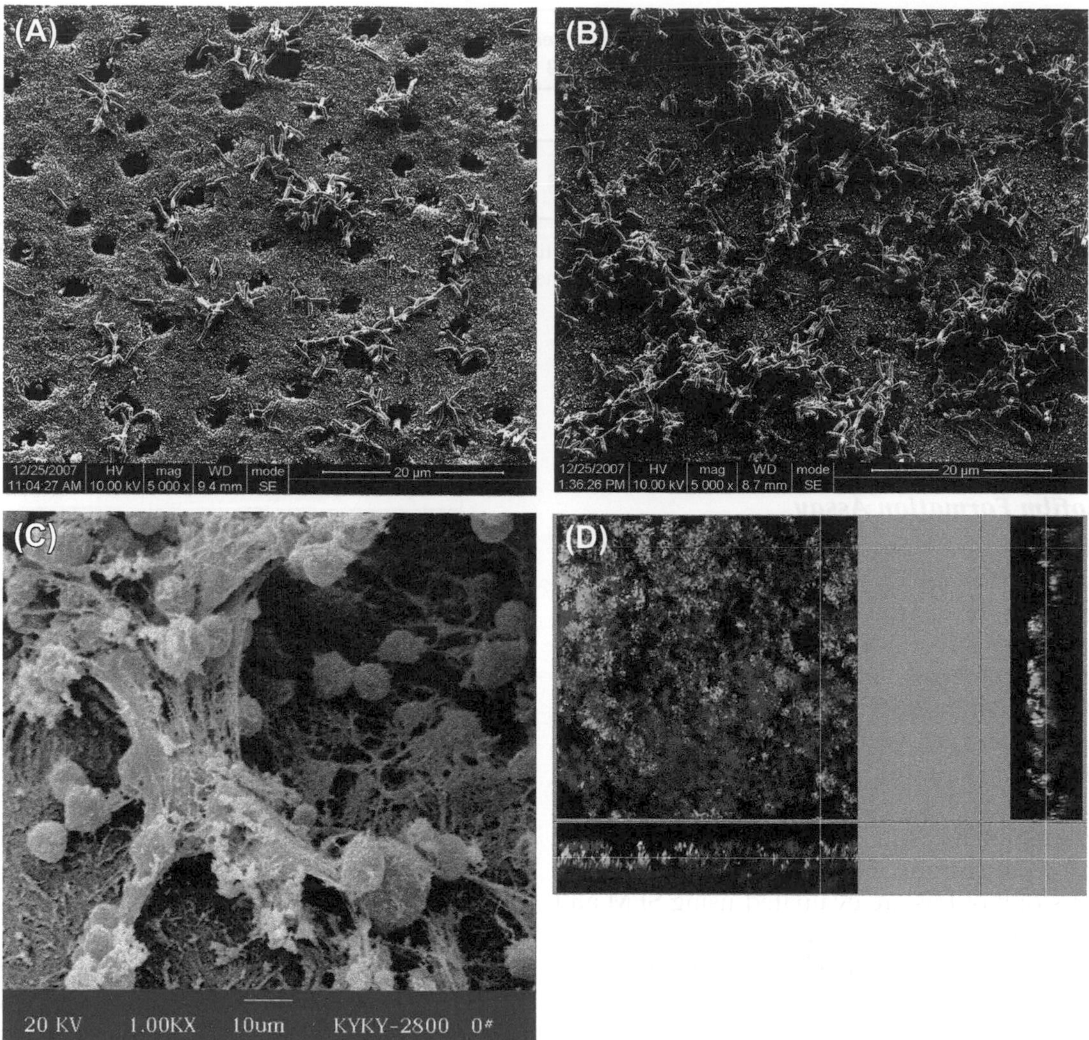

FIGURE 2.35 (A) Thirty-six-hour biofilm in dentin of *Actinomyces viscosus* (SEM). (B) Forty-eight-hour biofilm in dentin of *Actinobacillus actinomycetemcomitans* (SEM). (C) Biofilm from infected root canal (SEM). (D) Biofilm structure of *Streptococcus mutans*. Under a laser confocal microscope (green laser, 504–511 nm), biofilms appear as a multilayer superimposed image due to yellow fluorescent staining of bacterial colonies and the plaque biofilm structure. The three-dimensional structure of biofilm is mushroom-like, full of channels and pores. Against the dark background, microbial colonies appear as green fluorescence while dead bacteria show up as red fluorescence.

was introduced to microbial ecology by Muyzer et al. in 1993. Within a short period of time, this method has become widely used in the analysis of microbial diversity in various complex samples including samples from the oral cavity. In both DGGE and TGGE, DNA fragments of the same length but with different sequences can be separated. Separation is based on the reduced electrophoretic mobility of partially melted double-stranded DNA molecules in polyacrylamide gels that contain either a linear gradient of DNA denaturants (a mixture of formamide and urea) in the case of DGGE or a linear temperature gradient in the case of TGGE. Melting domains, i.e., stretches of base pairs with identical melting temperatures (Tm), lead to the melting of DNA fragments within discrete domains. Once a domain with the lowest Tm reaches its Tm at a particular position in the denaturing or temperature gradient gel, and that segment of the DNA double helix transitions to melted single-stranded DNA, migration of the DNA molecule will virtually stop. Sequence variations within these melting domains cause the melting temperatures to vary, and molecules with different sequences will stop migrating at different positions in the denaturing or temperature gradient gel, therefore becoming separated. DNA bands in DGGE and TGGE profiles can be visualized using ethidium bromide, SYBR Green I, or silver stain (Figure 2.36).

2.5.2 Sanger Sequencing

Sanger sequencing, also known as the chain termination method, is a technique for DNA sequencing based

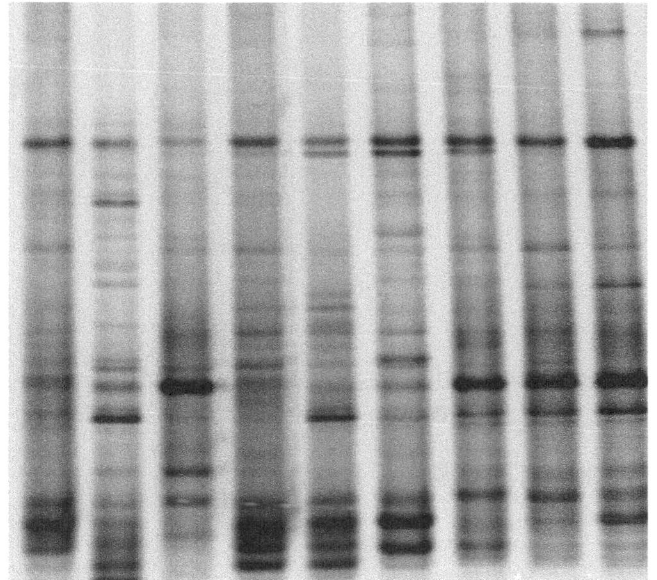

FIGURE 2.36 **Negative image of an ethidium bromide-stained DGGE gel loaded with 16S rRNA gene fragments.**

upon the selective incorporation of chain-terminating dideoxynucleotides (ddNTPs) by DNA polymerase during in vitro DNA replication. It was developed by Frederick Sanger and colleagues in 1977. It was the most widely used sequencing method for approximately 25 years before it was replaced by next-generation sequencing (NGS) methods.

Classical Sanger sequencing requires a single-stranded DNA template, a DNA polymerase, a DNA primer, normal deoxynucleosidetriphosphates (dNTPs), and modified nuclcotides (ddNTPs) that terminate DNA strand elongation. These ddNTPs lack a 3′-OH group that is required for the formation of a phosphodiester bond between two nucleotides, causing the extension of the DNA strand to stop when a ddNTP is added. The DNA sample is divided into four separate sequencing reactions, containing all four of the standard dNTPs (dATP, dGTP, dCTP, and dTTP), the DNA polymerase, and only one of the four ddNTPs (ddATP, ddGTP, ddCTP, or ddTTP) for each reaction. After rounds of template DNA extension, the DNA fragments that are formed are denatured and separated by size using gel electrophoresis with each of the four reactions in one of four separated lanes. The DNA bands can then be visualized by UV light or autoradiography, and the DNA sequence can be directly read off the gel image or the X-ray film (Figure 2.37). The ddNTPs may also be radioactively or fluorescently labeled for detection in automated sequencing machines. The four reactions can be incorporated into one reaction run, and the DNA sequence can be read from radioactive or fluorescent labels.

2.5.3 Next-generation Sequencing

Sanger sequencing enabled scientists to elucidate genetic information from a variety of biological systems. However, wide use of this technology has been hampered due to inherent limitations of throughput, scalability, speed, and resolution. Next-generation sequencing, also known as massively parallel sequencing or deep sequencing, has been developed to overcome these barriers.

In principle, NGS technology is based on sequentially identifying bases in a small fragment of DNA using emitted signals while each fragment is resynthesized from a DNA template strand. NGS proceeds in a massively parallel fashion, which enables rapid sequencing of large stretches of DNA spanning entire genomes.

2.5.3.1 *Pyrosequencing*

Pyrosequencing is a method of DNA sequencing based on the "sequencing by synthesis" principle. It relies on the detection of pyrophosphate release along with nucleotide incorporation. Sequences in each sample are tagged with a unique barcode either by ligation or by using a barcoded primer when amplifying each sample by PCR. The method amplifies DNA inside water droplets in an oil solution (emulsion PCR), and each droplet contains a single DNA template attached to a single primer-coated bead. The sequencing machine contains many picoliter-volume wells, each containing sequencing enzymes and a single bead. The sequence of the single-stranded DNA can be determined by the light emitted upon incorporation of the complementary nucleotide because only one of four of the possible A/T/C/G nucleotides can complement the DNA template. The technique uses luciferase to generate light for detection of the individual nucleotides, and the combined data are used to generate the template sequence.

Pyrosequencing was commonly used for genome sequencing or resequencing during the last decade. It was widely used in the analysis of the oral microbiome. However, a limitation of the method is that the length of individual DNA reads is approximately 300–500 nucleotides, shorter than the 800–1000 that can be obtained using Sanger sequencing. This can make the process of genome assembly more difficult, particularly for sequences containing a large amount of repetitive DNA (Figure 2.38).

2.5.3.2 *Illumina Sequencing*

Illumina sequencing is based on the incorporation of reversible dye terminators that enable the identification of single bases as they are incorporated into DNA strands. The basic procedure is as follows. DNA molecules are first attached to primers on a slide and amplified so that local clusters are formed. The four types (A/T/C/G) of reversible terminating nucleotides are added, and each nucleotide is fluorescently labeled with a different color

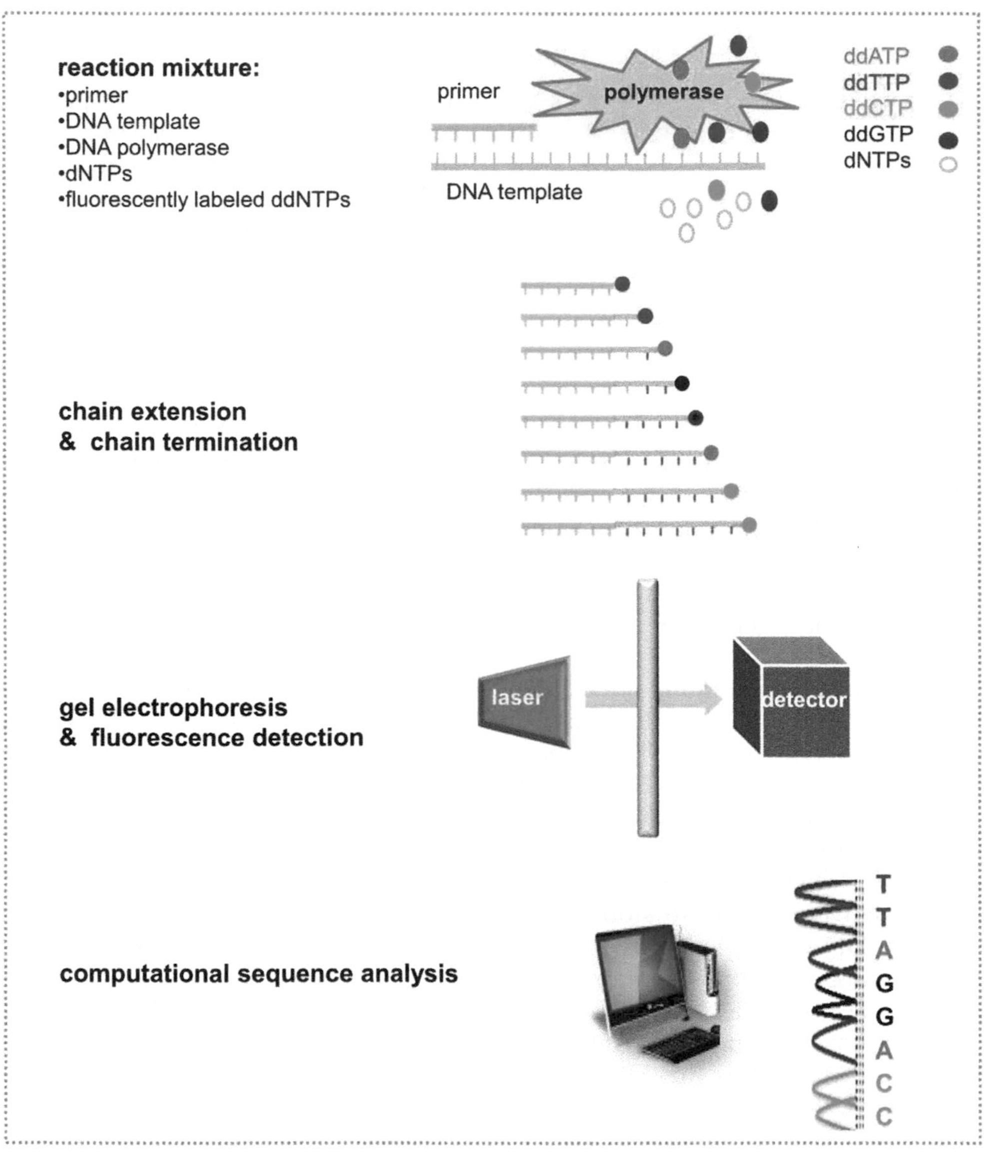

FIGURE 2.37 **Sanger sequencing.**

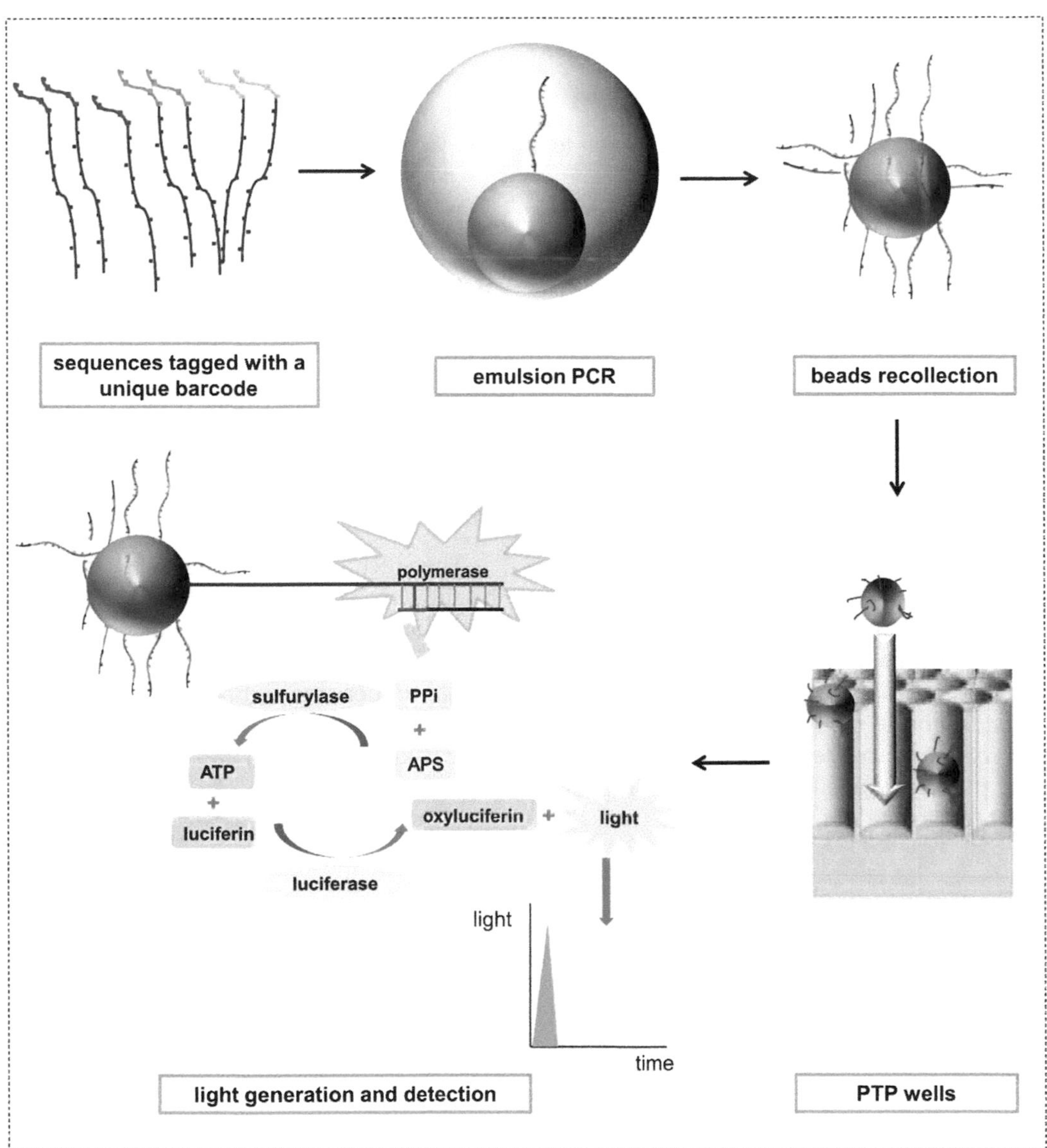

FIGURE 2.38 **Pyrosequencing.**

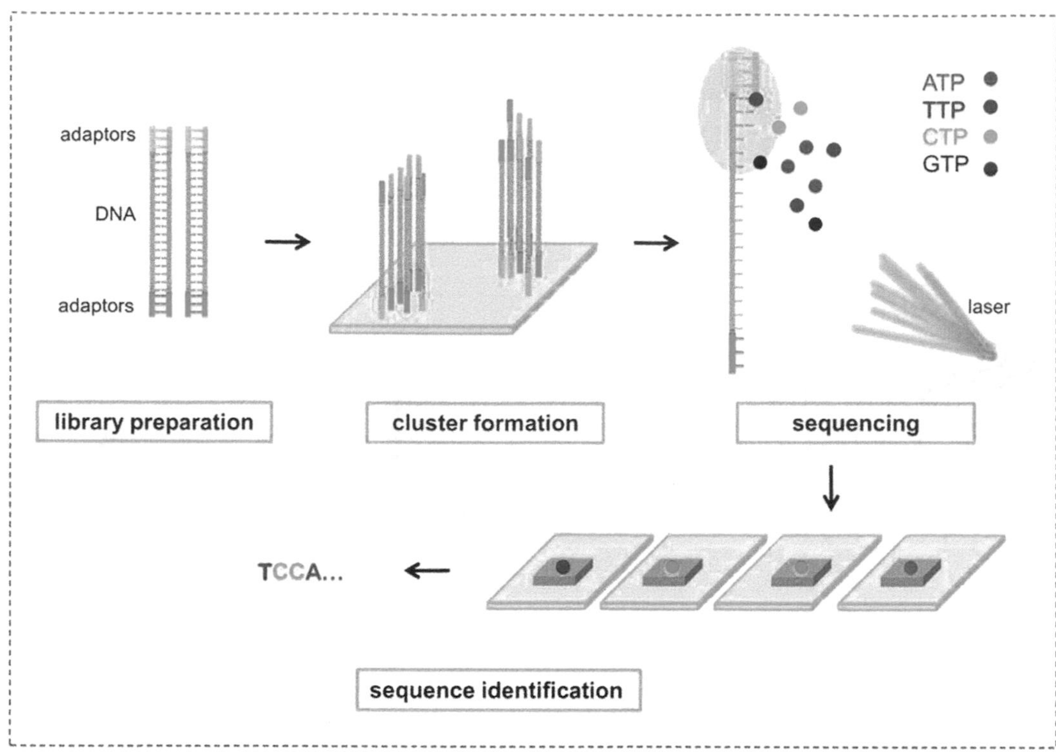

FIGURE 2.39 **Illumina sequencing.**

and attached to a blocking group. The four nucleotides then compete for binding sites on the template DNA to be sequenced, and nonincorporated molecules are washed away. After each synthesis, a laser is applied to remove the blocking group and the probe. A detectable fluorescent color specific to one of the four bases then becomes visible, allowing for sequence identification and initiating the beginning of the next cycle. The process is repeated until the entire DNA molecule is sequenced.

This technique offers some advantages over traditional sequencing methods such as Sanger sequencing. The automated nature of Illumina sequencing makes it possible to sequence multiple strands at once and obtain actual sequencing data quickly. In addition, this method only utilizes DNA polymerase in contrast with multiple, expensive enzymes required by pyrosequencing (Figure 2.39).

CHAPTER

3

Supragingival Microbes

3.1 GRAM-POSITIVE BACTERIA

3.1.1 *Actinomyces*

Actinomyces are irregular gram-positive bacilli and are also commonly found anaerobic bacteria in oral samples. When they were first discovered, actinomycetes were believed to be fungi or were grouped as "other microorganism." Recently, numerous studies have shown that actinomycetes have general characteristics common to bacteria and were classified as prokaryotic organisms. In the 1984 edition of *Bergey's Manual of Determinative Bacteriology*, *Actinomyces* were included in the group of gram-positive irregular bacilli. Common members of the *Actinomyces* genus in oral microbiology are *Actinomyces israelii*, *Actinomyces naeslundii*, *Actinomyces odontolyticus*, and *Actinomyces viscosus*. They are characterized by a relatively high GC content in their DNA, ranging from 57% to 69% (Tm method). The type species of this genus is *Actinomyces bovis*.

The shape and size of bacterial cells can vary greatly but are often found to be irregular branched bacilli with a diameter of 0.2–1.0 μm and a length of about 5.0–10.0 μm. The cells are usually rod-shaped, but can occasionally be club-shaped with irregular arrangements including single, paired, chain, clusters, and fence-shaped. The cells produce no spores, show no motility, and also do not produce conidia. The main distinguishing characteristic of *Actinomyces* cells is that the cell wall does not contain DAP and glycine.

There are differences at the species level regarding oxygen sensitivity, but a primary culture of *Actinomyces* requires an anaerobic environment. Spider-like microcolonies or branched mycelia can form on agar plates after 18–24 h incubation. These typical spider web-shaped colonies can help identify the genus of bacteria.

All strains of actinomycetes can ferment glucose and fructose to produce acid, without producing gas. Other tests, including fermentation of raffinose, xylose, cellobiose, laetrile, ribose, salicin, catalase production, reduction test using nitrate and nitrite, urea hydrolysis, or gelatin, can help identify different species.

Actinomycetes are normal members of the oral flora and are the dominant bacteria in dental plaque. *A. israelii*, *A. naeslundii*, *A. odontolyticus*, and *A. viscosus* can be detected in human dental plaque, dental calculus, and saliva. The main colonization site of *Actinomyces mai* is in the gingival sulcus. Clinical and epidemiological investigations indicate that *A. israelii* can cause actinomycosis, conjunctivitis, lachrymal and other diseases of the face, neck, lung, and abdomen. *A. naeslundii* and *A. mai* can be detected in clinical samples of gingivitis, periodontitis, pulp periapical infection, and pericoronitis. *A. viscosus* is suspected to be a cariogenic bacterium.

3.1.1.1 Actinomyces israelii

A. israelii are gram-positive irregular bacilli (Figure 3.1(A)–(C)). The culture atmosphere requires CO_2, as the culture grows poorly or not at all under ordinary atmospheric environment. Bacterial growth can be inhibited by 4–6% NaCl, 20% bile, or 0.005% crystal violet. Characteristic colonies are shown in Figure 3.1(D)–(F). The DNA GC content ranges from 57% to 65% when analyzed using the Tm method. The type strain is ATCC12102 (WVU46, CDCX523, W855).

The main habitat of this species is the oral cavity. *A. israelii* often colonizes the tonsils and plaque but can also be detected in the human gut and the female reproductive tract. This bacterium is mainly a pathogen of the face and neck, and causes lung and abdominal actinomycosis. It can also infect the lachrymal sac and conjunctiva. It is detected in oral mixed infections such as gingivitis, periodontitis, and pericoronitis, but the link between *A. israelii* and the infections is unclear.

Young colonies of *A. israelii* are typical filamentous microcolonies. Mature colonies are characteristically less than 2 mm in diameter, raised, white, opaque, and molar-shaped. The species shows no β-hemolytic reaction on blood agar.

Atlas of Oral Microbiology
http://dx.doi.org/10.1016/B978-0-12-802234-4.00003-3

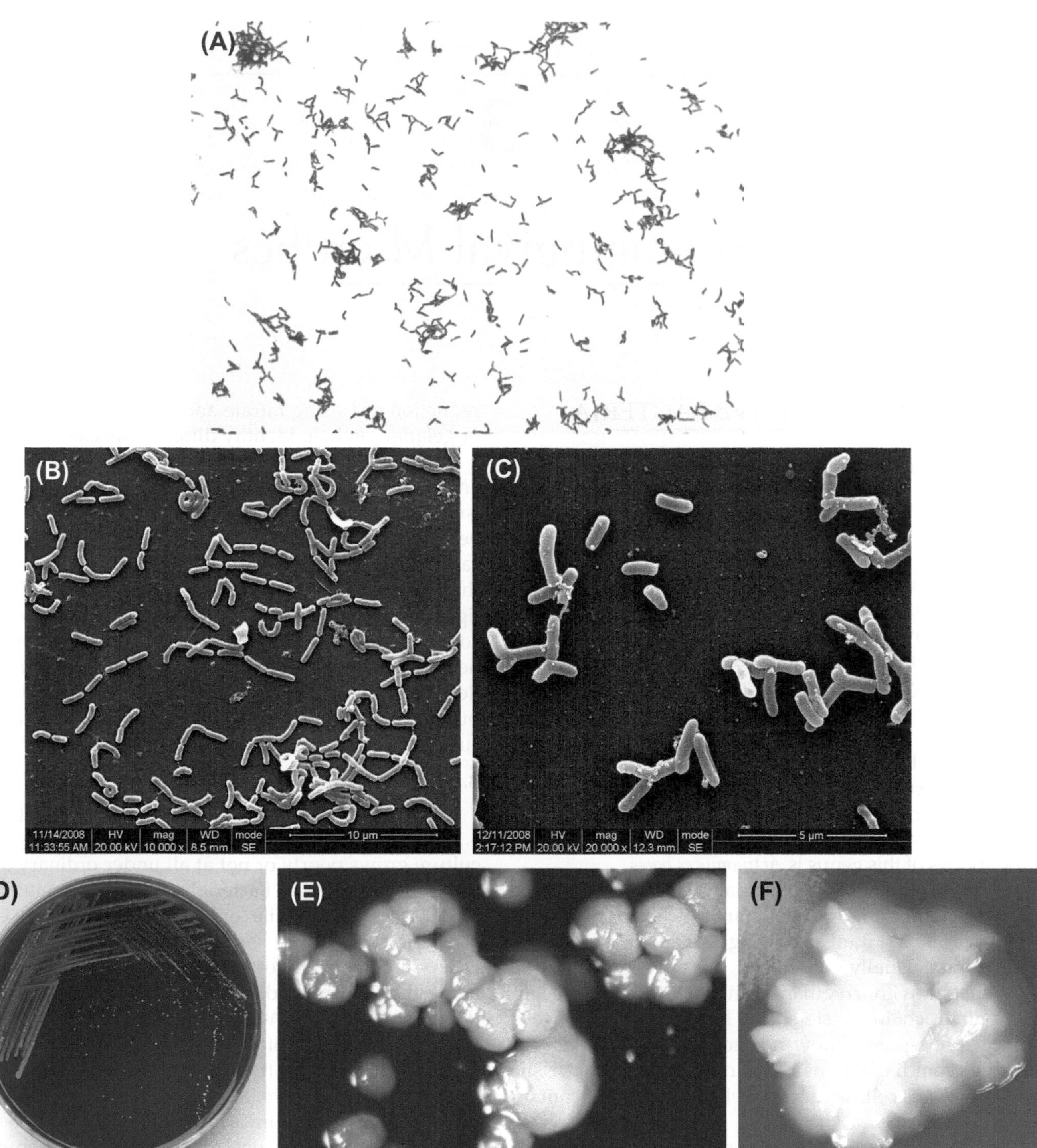

FIGURE 3.1 (A) *Actinomyces israelii* cells (Gram stain). (B) *A. israelii* cells (SEM). (C) *A. israelii* branch cells (SEM). *A. israelii* always display branching rods and short filamentous with no spores, Gram stain negative. (D) *A. israelii* colonies (BHI blood agar). (E) *A. israelii* molar-shaped colonies (stereomicroscope). (F) *A. israelii* colonies detached from plaque samples (stereomicroscope).

3.1.1.2 Actinomyces naeslundii

A. naeslundii is a gram-positive irregular bacillus (Figure 3.2(A) and (B)). Under aerobic conditions without CO_2, the culture may not grow. However, cultures can be grown in an anaerobic environment without CO_2. Some strains can be grown at 45 °C. Growth can be inhibited by 6% NaCl. Characteristic colonies are shown in Figure 3.2(C) and (D), and broth culture is shown in Figure 3.2(E). The DNA GC content is approximately 63–69% using the Tm method.

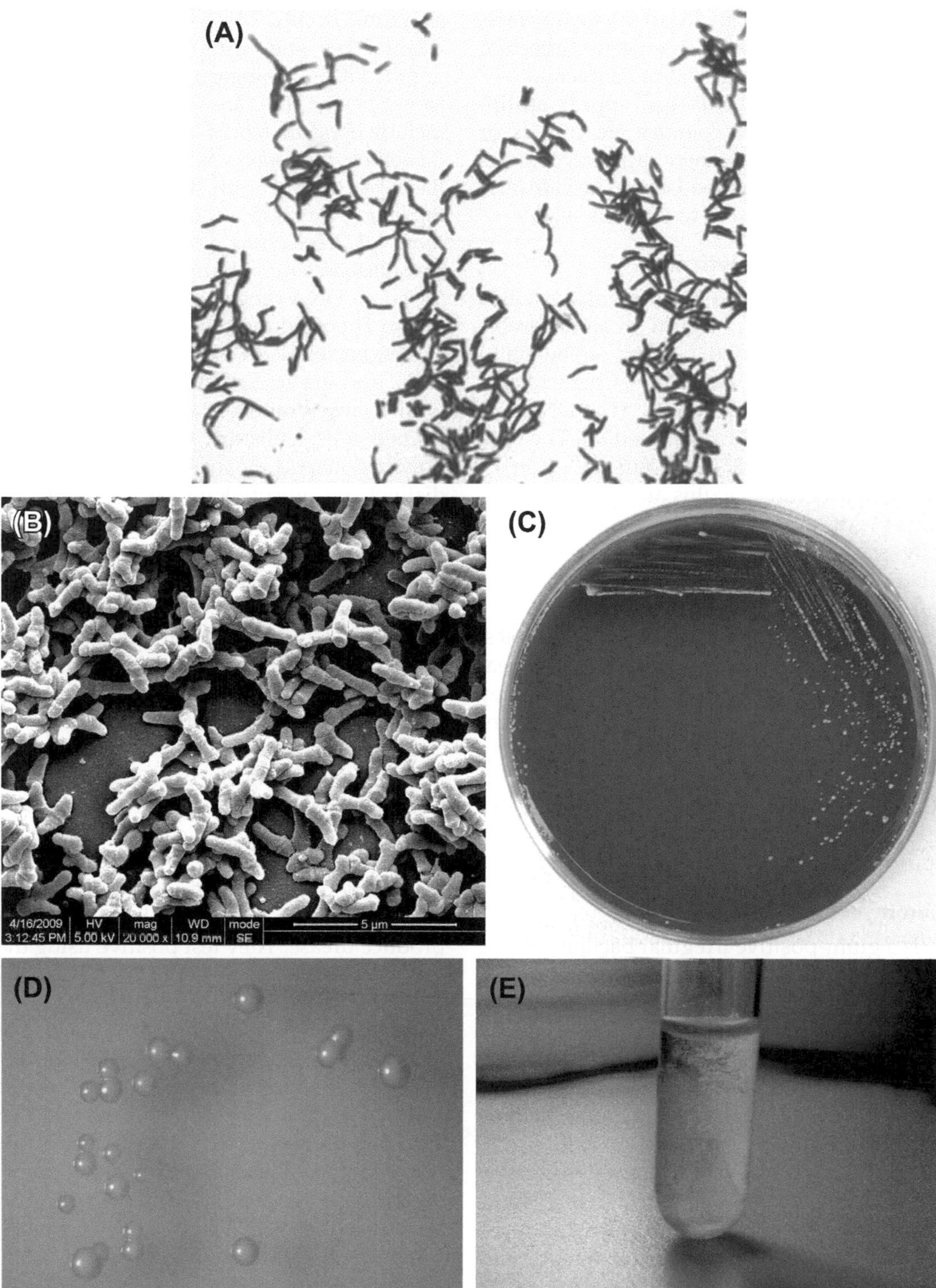

FIGURE 3.2 (A) *Actinomyces naeslundii* cells (Gram stain). (B) *A. naeslundii* cells (SEM). *A. naeslundii* cells are mainly irregular branched rods or short filaments without spores. The cells are gram-negative. (C) *A. naeslundii* colonies (BHI blood agar). (D) *A. naeslundii* colonies (stereomicroscope). (E) *A. naeslundii* liquid culture (BHI broth).

The type strain is ATCC12104 (NCTC10301, WVU45, CDCX454).

A. naeslundii is a member of the normal human oral flora and can be found on the tonsils and in dental plaque. It can take part in mixed bacteria infections and is one of the pathogenic bacteria in root caries. It is often detected in clinical specimens of periodontitis or infected root canals, but the pathogenesis is unclear. This bacterium can lead to actinomycosis at many locations such as face, neck, chest, abdomen, and eye, and can cause infection of the female genital tract and knee joint empyema.

Young colonies of *A. naeslundii* can appear as filiformed microcolonies. Mature colonies are convex or flat, rough or smooth without center sag. The colonies show no hemolytic reaction on blood agar. The culture is muddy in broth culture and sticks to the flask wall.

3.1.1.3 Actinomyces odontolyticus

A. odontolyticus is a gram-positive irregular bacillus (Figure 3.3(A) and (B)). It cannot grow in an anaerobic environment unless CO_2 is added. Growth can be inhibited by 5–20% bile or 0.005% crystal violet. Characteristic colonies are shown in Figure 3.3(C)–(E). The DNA GC content is 62% when analyzed by Tm method. The type strain is ATCC17929 (NCTC9935, WVU867, CDCX363).

Dental plaque and dental calculus are the main habitats. This species is often involved in eye infections and is occasionally seen in advanced actinomycosis. The relationship between *A. odontolyticus* and periodontosis and dental caries remains to be confirmed.

A. odontolyticus can generate filiformed microcolonies. The mature colonies on the surface of BHI agar measures are less than or equal to 2mm in diameter. They are white or gray-white, opaque, and do not show a central sag or filamentous edge. The important identifying feature of this species is that the colony turns dark red after 2 days of culture on the surface of blood agar under an anaerobic environment. The color clears after the cells are placed at room temperature.

3.1.1.4 Actinomyces viscosus

A. viscosus is a gram-positive irregular bacillus (Figure 3.4(A)–(C)). Primary cultures grow under anaerobic conditions and CO_2 can stimulate its growth. Subcultures can grow in common atmospheric conditions. Some cells can grow at temperatures up to 45 °C. Characteristic colonies are shown in Figure 3.4(D) and (E). The DNA GC content ranges from 59% to 70% using the Tm method. The type strain is ATCC15987 (WVU745, CDCX603, A828).

Humans' and other warm-blooded animals' oral cavity, including subgingival plaque, transparent plaque, and dental calculus, are the main sites of growth. *A. viscosus* is one of the cariogenic bacteria found in root caries and is related to periapical infections, dacryosolenitis, and abdominal and faciocervical actinomycosis.

A. viscosus can generate filiformed microcolony. The mature colony does not have center sag. The typical colony is big and sticky.

3.1.2 *Bifidobacterium*

Bifidobacterium is a genus of bacteria with various forms; nonmotile they are gram-positive, nonsporulating, anaerobic bacilli. These bacteria were first isolated from infant feces and attracted attention because of their important physiological significance to the host organism. Species that are important human gut bacteria include *Bifidobacterium bifidum*, *Bifidobacterium infantis*, *Bifidobacterium adolescentis*, and *Bifidobacterium longum*. Bacteria isolated from the oral cavity belonging to the *Bifidobacterium* spp. include mainly *Bifidobacterium dentium*, *Bifidobacterium breve*, *Bifidobacterium inopinatum*, and *Bifidobacterium denticolenu*. The percent GC in *Bifidobacterium* DNA ranges from 55% to 67% when analyzed by the Tm or Bd method. The type species is *B. bifidum*.

The bacterial cells are short and thin, with pointed ends, and are irregular. They also appear as long cells with many branches and slightly branching spoon-shaped cells. Cells are arranged as single cells, chains, polymer-shaped, V-shaped, or palisade-shaped. Their distinct cell morphology can be helpful in differentiating bacteria belonging to this genus. For example, *B. bifidum* appear as flask-shaped cells, while *Bifidobacterium asteroides* are star-shaped. All members of this genus are gram-positive.

Bifidobacterium are anaerobes, and most strains cannot grow under 90% air and 10% CO_2. Colonies formed on agar plates are convex, creamy or white, glossy, smooth, neat-edged, sticky, and soft.

The main terminal acid products in liquid culture medium containing glucose are acetic acid and lactic acid, but a few species also make formic acid and succinic acid. However, no butyric acid or propionic acid is formed, and no CO_2 is generated (with the exception of gluconate degradation). *Bifidobacterium* can ferment carbohydrates to produce acid, and generally do not reduce nitrate or produce urease. They test positive using the catalase test.

3.1.2.1 Bifidobacterium dentium

Originally, *B. dentium* was isolated from pus specimens and named *B. appendicitis*. Later, researchers isolated similar bacteria from the adult dental caries, feces, and vagina, and they were then named *Actinomyces eriksonii* or grouped into *B. adolescentis*. According to later research, these bacteria make up an independent branch on the phylogenetic tree, and the species was named *B. dentium* in the 1970s. The percent GC in their DNA is 61% (Tm method) and type species is ATCC27534 (Reference strains B764).

The cells are anaerobic gram-positive irregular bacilli (Figure 3.5(A)). Some strains are resistant to oxygen in the presence of CO_2. The optimal temperature for the growth of this bacterium is 37–41 °C, and the optimal pH value ranges from 6.5 to 7.0. TPY culture medium supplemented with neomycin, kanamycin, and various salt solutions is commonly used as the selective culture medium. Characteristic colonies are shown in Figure 3.5(B)–(D).

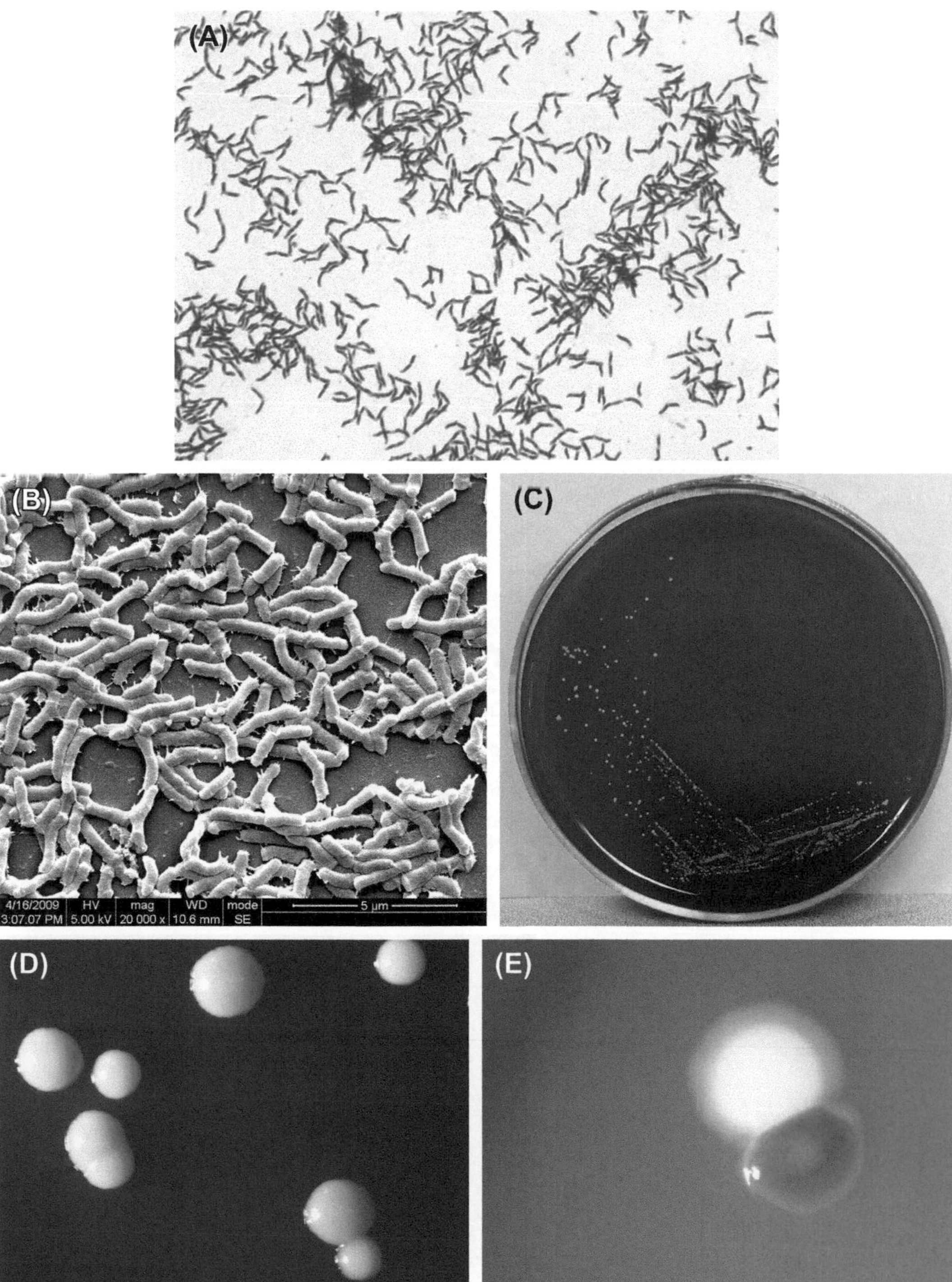

FIGURE 3.3 (A) *Actinomyces odontolyticus* cells (Gram stain). (B) *A. odontolyticus* cells (SEM). *A. odontolyticus* cells are mainly irregularly rod-shaped. Ball-like rods, branched rods, or filiform cells can also be observed. The cells are gram-positive. (C) *A. odontolyticus* colonies (BHI blood agar). (D) Red colonies of *A. odontolyticus* (stereomicroscope). (E) *A. odontolyticus* colonies in dental plaque (stereomicroscope).

B. dentium is biochemically active. It can ferment D-ribose, L-arabinose, lactose, sucrose, cellobiose, trehalose, raffinose, melibiose, mannitol, salicin, starch, galactose, maltose, fructose, xylose, mannose, and glucose to produce acid, but cannot ferment sorbitol and inulin. It cannot reduce nitrate and tests negative for both the urease test and the catalase test.

The distribution of the bacteria in the oral cavity and its pathogenic mechanism are not clearly characterized.

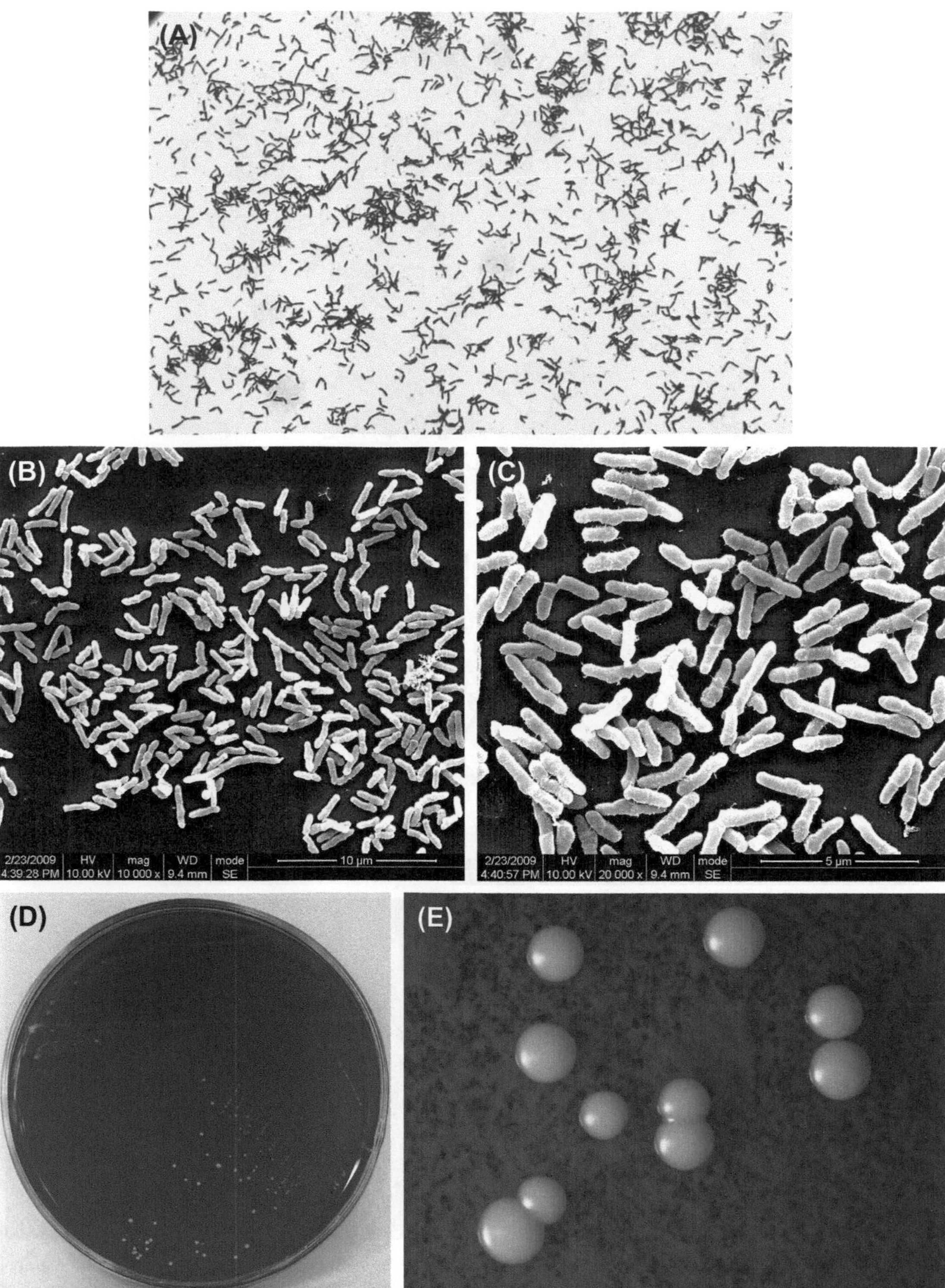

FIGURE 3.4 (A) *Actinomyces viscosus* cells (Gram stain). (B) *A. viscosus* cells (SEM). (C) *A. viscosus* cells (SEM). *A. viscosus* cells are mainly irregular short or moderate-length rod-shaped without spores. Branch rods or short filiform also can be seen, Gram stain is positive. (D) *A. viscosus* colonies (BHI blood agar). (E) *A. viscosus* colonies (stereomicroscope).

B. dentium forms colonies that can be described as spherical, lustrous, smooth, convex, gray-white, sticky and soft when plated on BHI blood agar and *Bifidobacterium* selective culture medium.

3.1.2.2 Bifidobacterium breve

B. breve is an anaerobic gram-positive irregular bacillus (Figure 3.6(A)–(C)). Its culture characteristics are similar to those of *B. dentium*, and characteristic colonies

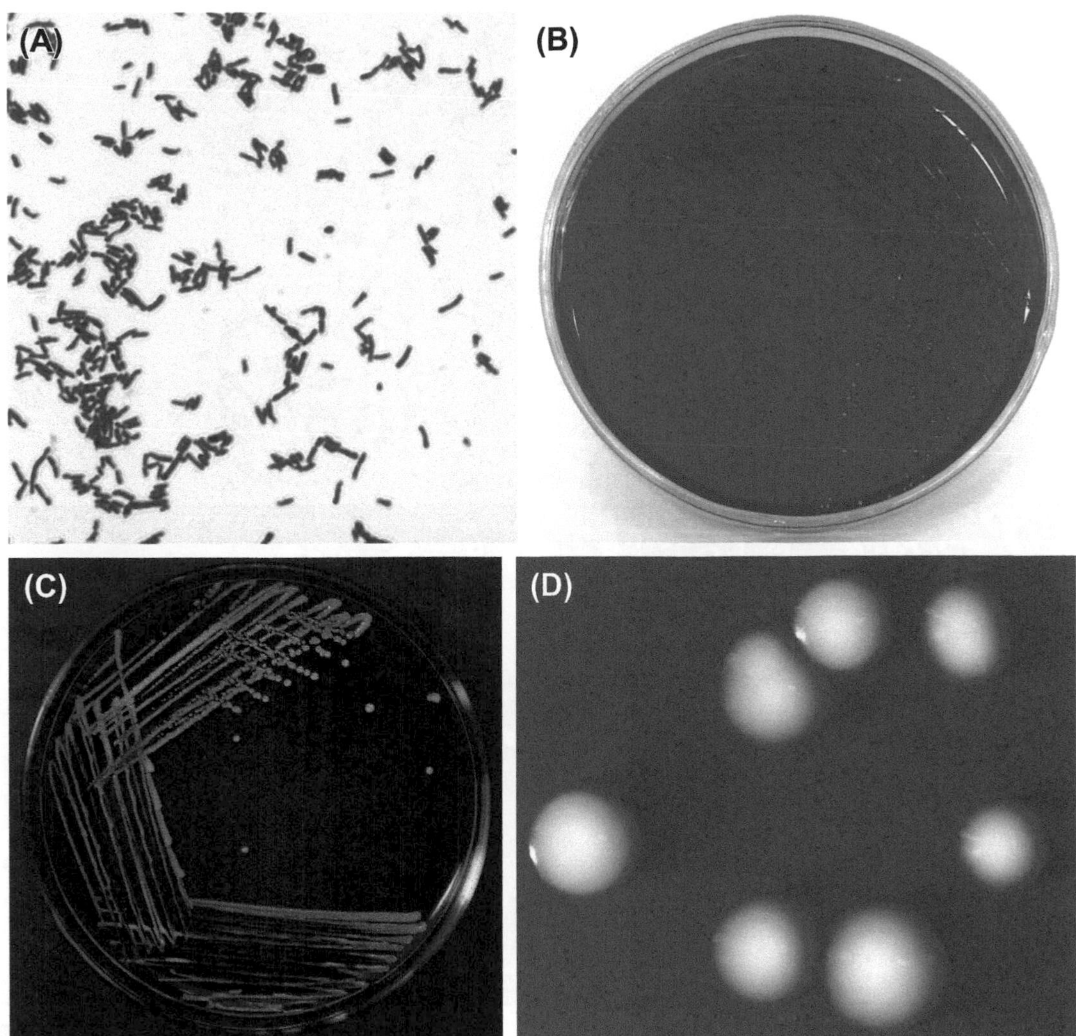

FIGURE 3.5 (A) *Bifidobacterium dentium* cells (Gram stain). (B) *B. dentium* colonies (BHI blood agar). (C) *B. dentium* colonies (*B. dentium* selective agar). (D) *B. dentium* colonies (stereomicroscope).

are shown in Figure 3.6(D) and (E). The percent GC in *B. breve* DNA is 58% by the Tm method and the type species is ATCC15700.

B. breve can ferment D-ribose, lactose and raffinose but does not ferment L-arabinose and starch.

B. breve on agar plates form colonies that are spherical, smooth, translucent, gray-white, sticky, and soft.

3.1.3 *Lactobacillus*

Lactobacillus is a group of anaerobic or microaerobic gram-positive bacilli that do not produce spores. Bacteria of this genus form part of the normal flora of the human oral cavity and intestinal tract. This genus is cariogenic, as they are detected in decayed oral cavity materials. This genus includes 44 species according to *Bergey's Manual of Systematic Bacteriology* and also contains seven subspecies. The common species found in the oral cavity include *Lactobacillus acidophilus*, *Lactobacillus salivavius*, *Lactobacillus plantarum*, *Lactobacillus fermentum*, *Lactobacillus brevis*, and *Lactobacillus casei*. The percentage GC in the DNA is 40% (by either Tm method or Bd method). The type species is German type *Lactobacillus*.

The shapes and sizes of the bacterial cells can vary greatly. They can be vimineous, stubbed, bent, bacilliform, clavate, club-shaped, etc. However, most *Lactobacillus* cells are fairly regular with no branching. The cells are square or obtuse at the ends when compared with other gram-positive nonsporulating bacilli. They produce no spores and no capsules and stain gram-positive.

Surface culture on a solid medium is best when performed in anaerobic or microaerophilic conditions. However, some species must be cultivated under anaerobic conditions. Some members of this genus can grow within the 15–45°C range, in the presence of 5–10% CO_2, which promotes bacterial growth. As acid-producing bacteria, low pH Rogosa agar is the culture medium of choice for many strains of *Lactobacillus*. The optimal pH for growth is 5.5–6.2.

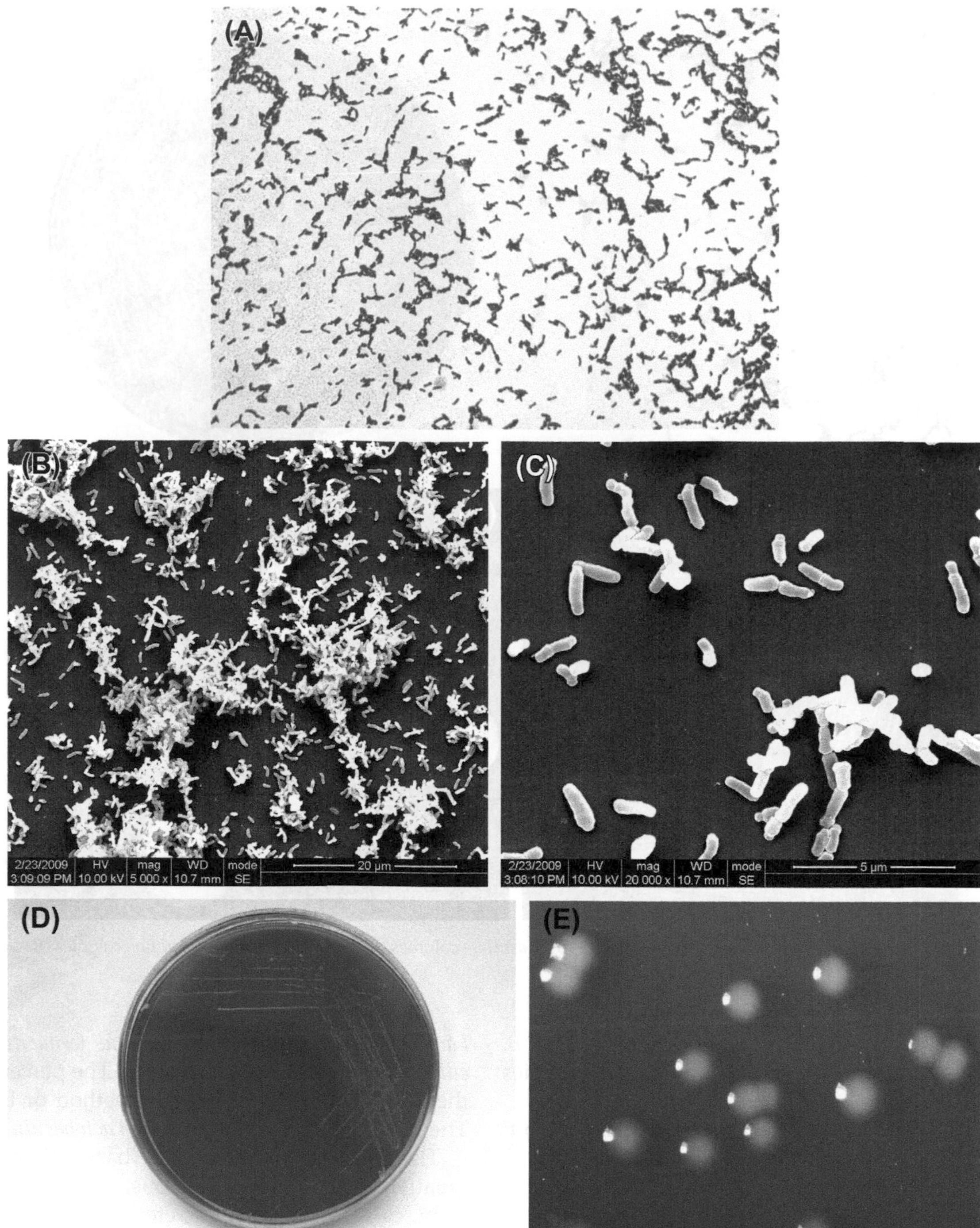

FIGURE 3.6 (A) *B. breve* cells (Gram stain). (B) *B. breve* cells (SEM). (C) *B. breve* cells (SEM) *B. breve* cells are thin and short bacilli. Nonsporulated, nonmotile, gram-positive. (D) *B. breve* colonies (BHI blood agar). (E) *B. breve* colonies (stereomicroscope).

The colonies are round, white or gray, transparent or nontransparent with a diameter from pinprick-sized to 2 mm on the agar surface. Smooth colonies are soft, raised, and lustrous, and the edge of the colony is neat. The surface of rough colonies is dry, flat, and lackluster, and the edge is not neat. The bacteria normally do not produce pigment.

Lactobacilli can ferment glucose to produce acid, and are negative for catalase, urease, and cytochrome enzyme. They do not produce benzpyrole, cannot reduce nitrate, cannot hydrolyze gelatin, and cannot produce H_2S. Sugar fermentation test and arginine hydrolysis test can help identify the genus of bacteria.

The bacteria can promote the development of tooth decay, as its detection is significantly increased in deep caries material.

3.1.3.1 Lactobacillus acidophilus

These are gram-positive regularly shaped bacteria (Figure 3.7(A) and (B)) that grow well under anaerobic conditions. Cells grow well at 45 °C but do not grow at 15 °C. BHI blood agar and Rogosa agar are the commonly used media, and the latter is a selective medium. Characteristic colonies are shown in Figure 3.7(C)–(G).

L. acidophilus is obligate homofermentative bacteria. They can produce D- or L-lactic acid, ferment glucose, and most strains can also ferment starch. *L. acidophilus* cannot produce ammonia from arginine. The GC content

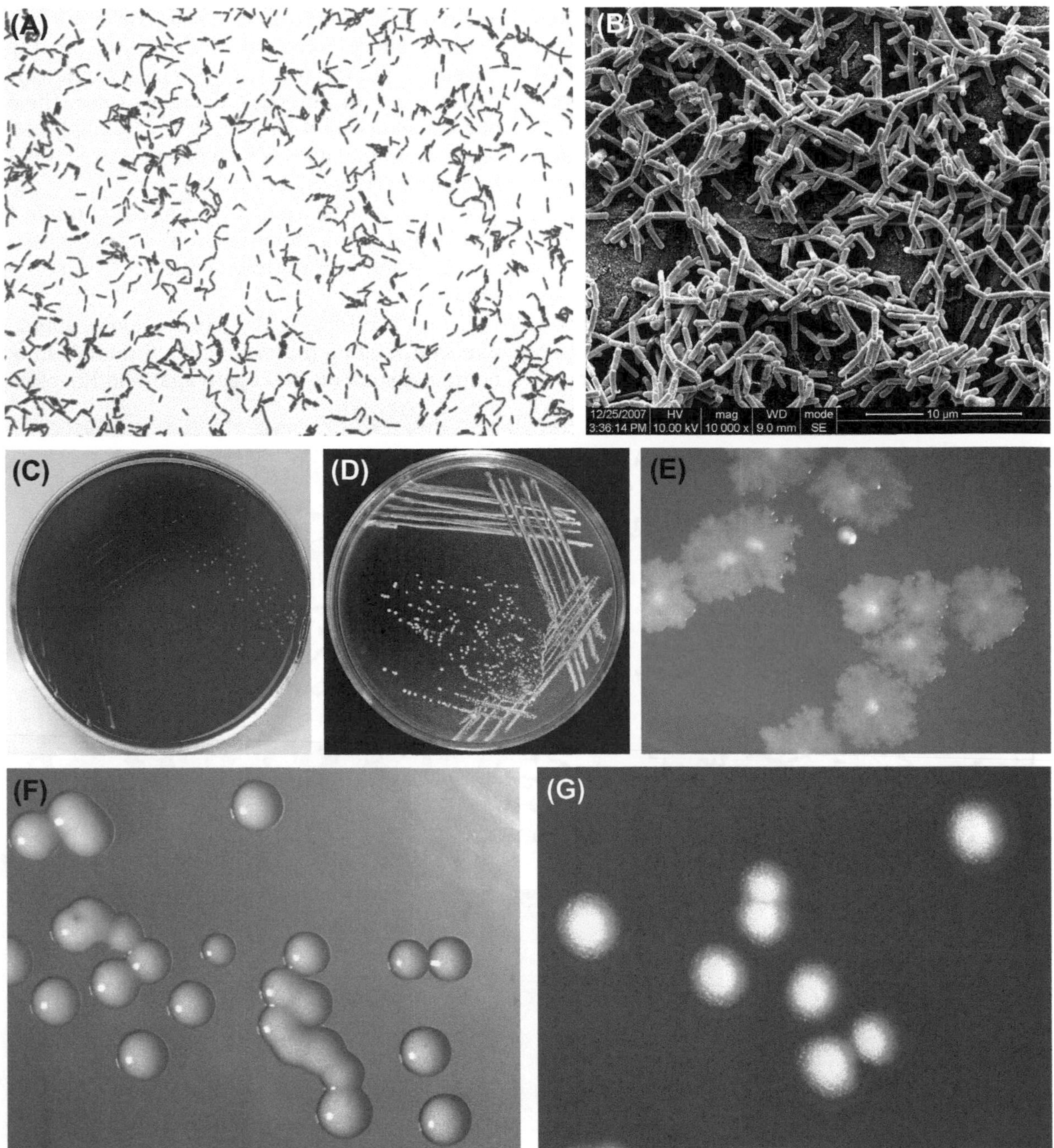

FIGURE 3.7 (A) *Lactobacillus acidophilus* cell (Gram stain). (B) *L. acidophilus* cell (SEM). Cells of *L. acidophilus* measure 0.3–0.4 μm × 1.5–4.0 μm in size. They are arranged as single cells, in pairs, or as short chains. They have rounded ends and no muramic acid in the cell wall. Cells stain gram-positive. (C) Colonies of *L. acidophilus* (BHI blood agar). (D) Colonies of *L. acidophilus* (Rogosa agar). (E) Rough colonies of *L. acidophilus* (stereomicroscope). (F) Smooth colonies of *L. acidophilus* (stereomicroscope). (G) Surface features of the *L. acidophilus* rough colonies (stereomicroscope).

in its genomic DNA is 32–37% (by Tm method or Bd method). The type strain is ATCC4356.

L. acidophilus is mainly isolated from the gastrointestinal tract of humans and animals, human mouths, and the human vagina. *L. acidophilus* can be isolated from a minority of neonates' mouths. As the children grow older, the number of bacteria found in their mouth decreases gradually, until at age 2, only a very small amount of *L. acidophilus* can be detected. The main site of colonization of *L. acidophilus* in the mouth is dental plaque; it is relatively rare in saliva, on the tongue, or in the gingival sulcus. As it is often found in material from deep caries, *L. acidophilus* is believed to be associated with the development of dental caries.

L. acidophilus colonies can be separated into rough and smooth type. Hair-like structures can be observed under the stereomicroscope. Colonies do not produce pigment.

3.1.3.2 Lactobacillus casei

L. casei was originally divided into four subspecies: *L. casei* subsp. *casei*, *L. casei* subsp. *pseudoplantarum* (now known as *Lactobacillus paracasei* subsp. *paracasei*), *L. casei* subsp. *rhamnosas* (now known as *L. rhamnosus*), and *L. casei* subsp. *toleons* (now known as *L. paracasei* subsp. *tolerans*). A newly added subspecies is *L. casei* subsp. *alactosus* (now known as *L. paracasei* subsp. *paracasei*).

The cells stain gram-positive (Figure 3.8(A) and (B)). Cultures grow well under anaerobic conditions. BHI blood agar and Rogosa agar are the commonly used media, while the latter is a selective medium. Characteristic colonies are shown in Figure 3.8(C)–(E). The percent GC in its genomic DNA is 45–47% (Bd method). The type strain is ATCC393 (*L. casei* subsp. *casei*).

The main colonization sites of *L. casei* are the human intestine, mouth, and vagina. The bacteria can also be detected in milk and other dairy products. The main site of colonization in the mouth is in dental plaque. As it is often found in material from deep caries, it is believed to be a pathogen of dental caries.

L. casei is a facultative heterofermentative bacterial species. Other than the subspecies *L. casei* subsp. *rhamnosas*, other subspecies grow well at 15 °C but do not grow at 45 °C.

L. casei can form milky white colonies about 1 mm in diameter. Colony morphology is rounded, smooth, and nontransparent on BHI blood agar and Rogosa agar.

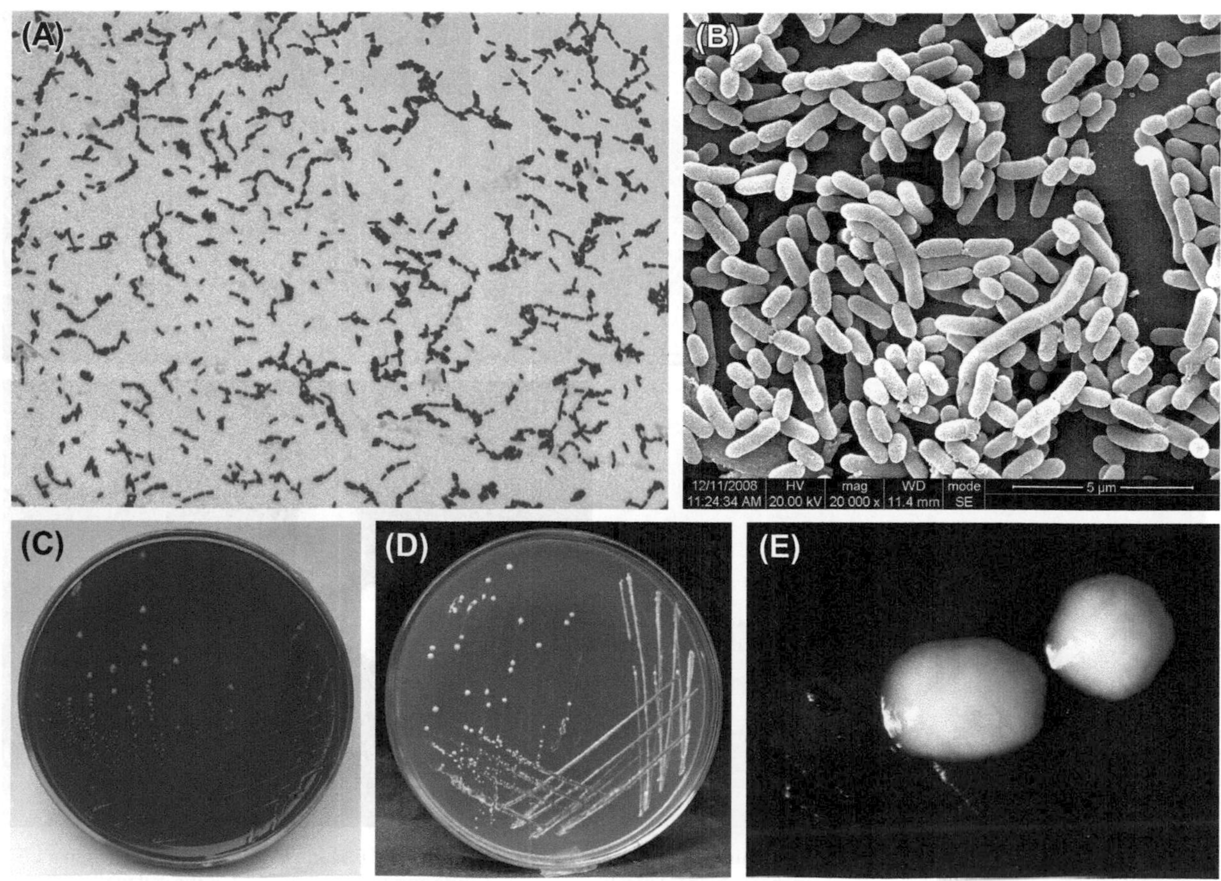

FIGURE 3.8 (A) *Lactobacillus casei* cell (Gram stain). (B) *L. casei* cell (SEM). *L. casei* cells measure 0.7–1.1 μm × 2.0–4.0 μm × 1.5–5.0 μm and are arranged in a chain. Individual cells usually have rounded ends. Cells stain gram-positive. (C) Colonies of *L. casei* (BHI blood agar). (D) Colonies of *L. casei* (Rogosa agar). (E) Cheese colonies of *L. casei* (stereomicroscope).

3.1.3.3 Lactobacillus fermentum

L. fermentum is a gram-positive bacterium (Figure 3.9(A)–(C)). Commonly used media to culture this species are BHI blood agar and Rogosa agar, where the latter is a selective medium. Characteristic colonies are shown in Figure 3.9(D)–(F).

Since this species can be detected in the human mouth and in yeast, dairy products, sourdough, and fermented

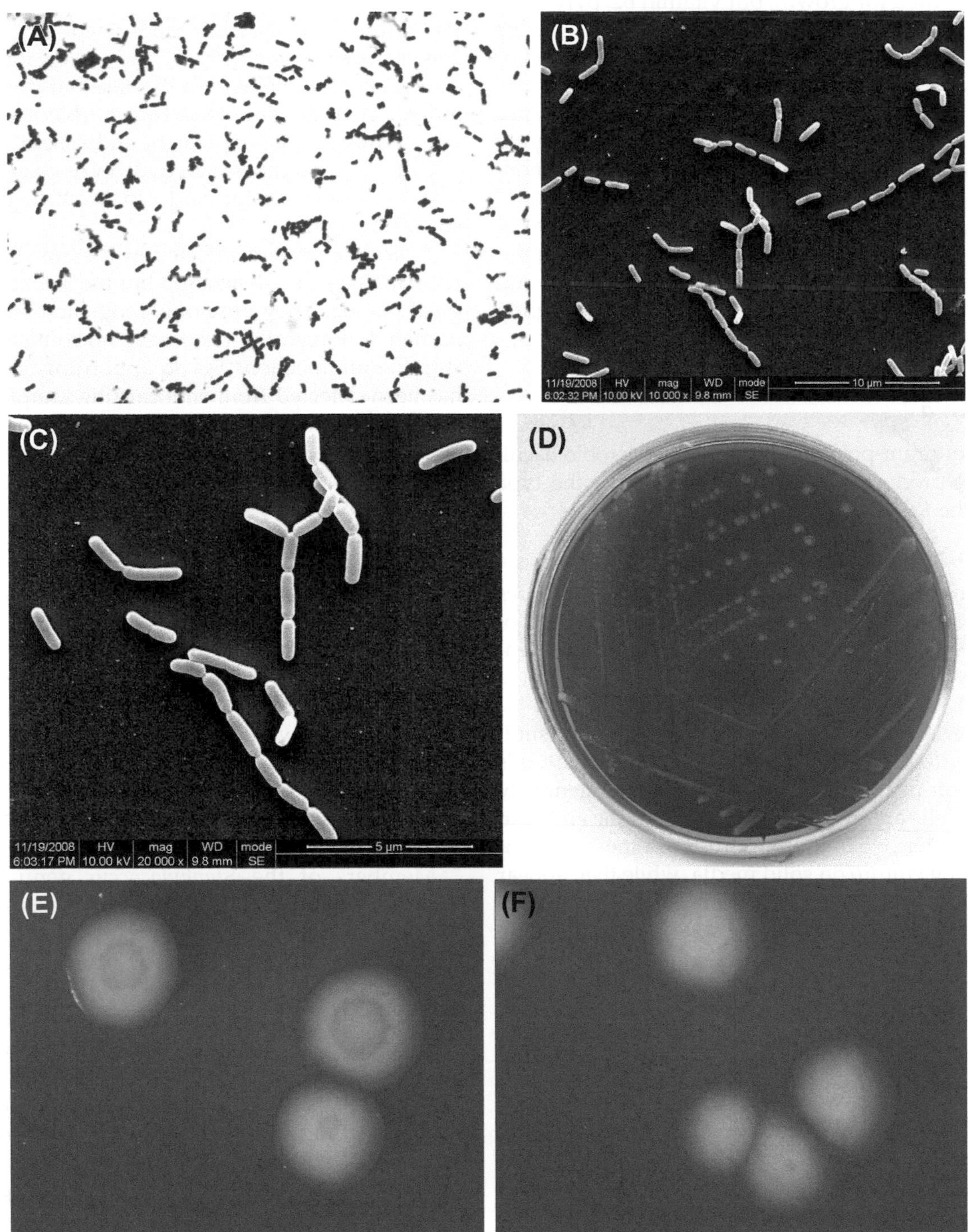

FIGURE 3.9 (A) *Lactobacillus fermentum* cell (Gram stain). (B) *L. fermentum* cell (SEM). (C) *L. fermentum* cell (SEM). The *L. fermentum* cell is 0.5–0.9 μm in diameter, but its length can vary quite significantly. The ends of the cells are square or obtuse. Most bacterial cells are arranged as a single cell or in pairs. They do not produce spores and are nonmotile. *L. fermentum* stains gram-positive. (D) Colonies of *L. fermentum* (BHI blood agar). (E) Concentric circle structure of *L. fermentum* colonies (stereomicroscope). (F) Surface of smooth colonies and rough, lined, myxoid colonies of *L. fermentum* (stereomicroscope).

plants, *L. fermentum* is considered to be related to the occurrence of oral infectious diseases such as dental caries and root canal infections.

L. fermentum usually grows well at 45 °C and does not grow at 15 °C. They are obligate heterofermentative bacteria. Calcium pantothenate, nicotinic acid, and thiamine are required for growth, but vitamin B2, pyridoxal, and folic acid are not required. The percent GC in its genomic DNA is 52–54% (analyzed by the Bd method or Tm method). The type strain is ATCC14931.

L. fermentum can form gray-white colonies about 1 mm in diameter. The shape of the colony is convex or slightly convex, the surface is smooth, and the colonies are non-transparent on BHI blood agar. A striking feature of *L. fermentum* colonies is the concentric circle structure at the center of the colony when observed under stereomicroscope. The smooth spherical colonies are convex and have neat edges, while rough myxoid colonies are slightly convex and have irregular edges and a granular surface.

3.1.4 *Rothia*

Rothia are gram-positive facultative anaerobic bacilli that do not produce spores. *R. dentocariose* is the type species of the *Rothia* genus.

3.1.4.1 Rothia dentocariose

R. dentocariose is a gram-positive bacillus that does not produce spores. The characteristic appearance of the cells is shown in Figure 3.10(A)–(C). The GC content of its DNA is 47–57% (Tm method). The type strain is ATCC17931.

The bacteria cells can be spherical, pleiomorphic (similar to *Corynebacterium diphtheria*), or filamentous. Cells stain as gram-negative. The cell diameter is generally 1.0 μm, but cells are irregular in shape, and the ends can reach a diameter of 5.0 μm. Cells appear almost filamentous following culture on solid media, while they appear spherical in broth media. Cells become almost completely spherical after growing for 2–3 days in stale broth media, but the coccoid morphology can be easily altered. *R. dentocariose* does not produce spores or a capsule. They are nonmotile and are not acid tolerant. Bacterial cells as viewed by SEM are shown in Figure 3.10(B) and (C).

These bacteria are facultative anaerobes. They grow well in an aerobic environment, although primary cultures require incubation under anaerobic conditions (80% N_2, 10% H_2, 10% CO_2). The optimal growth temperature is 35–37 °C. When cultured for 18–24 h under anaerobic conditions, young colonies are always filamentous and appear as spider-like colonies. When inoculated under aerobic conditions, young colonies can reach a diameter of 1 mm. The colony surface is smooth or grainy and often shows an umbrella edge. After 2 days culture, mature colonies can reach a diameter of 2–6 mm, with a milky, glossy, and smooth appearance (Figure 3.10(B)). Smooth colonies and rough colonies can coexist on the same agar plate, which may also show loose, crumbly, or sticky colonies. A handful of rough-type colonies may also take on a dry coil shape. Characteristic colonies are shown in Figure 3.10(D) and (E).

The main acid product is lactic acid, and *R. dentocariose* does not produce propionic acid when inoculated into PYG broth. *R. dentocariose* can ferment glucose, maltose, sucrose, trehalose, fructose, and salicylate to produce acid. Cells test positive for catalase but do not produce indole. They are able to reduce nitrate and nitrite and can hydrolyze esculin, starch, and casein. It is unclear whether *R. dentocariose* can hydrolyze gelatin. It is positive for urease activity and can produce H_2S in triple sugar iron agar.

R. dentocariose are detected in the human oral cavity. Its main sites of colonization are the saliva and subgingival plaque. They are nonpathogenic members of the human oral microflora and have no confirmed relationship to oral infections. As an opportunistic pathogen, it has been detected from endocarditis samples and other clinical infected specimens.

When inoculated under aerobic conditions, young colonies can reach a diameter of 1 mm. The colony surface is smooth or grainy and often presents an umbrella-like edge. After 2 days' culture, mature colonies can reach a diameter ranging from 2 to 6 mm, with a milky, glossy, and smooth appearance. Smooth colonies and rough colonies can exist simultaneously on the same agar plate, while colonies can also appear loose, crumbly, or sticky. A small number of rough-type colonies may also exhibit a dry coil shape. When cultured for 18–24 h under anaerobic conditions, young colonies are always filamentous and appear as spider-like colonies.

3.1.5 *Staphylococcus*

Members of the *Staphylococcus* genus are gram-positive cocci and belong to the *Micrococcus* family. The organisms are widely spread in the environment. Early on, three species were isolated from clinical samples: *Staphylococcus aureus*, *S. epidermidis*, and *S. saprophyticus*. In the early 1980s, analysis of biochemical reactions (e.g., mannitol fermentation) and cellular components (e.g., the availability of coagulase) resulted in the division of the *Staphylococcus* genus into subgroups of pathogenic and nonpathogenic species. In *Bergey's Manual of Systematic Bacteriology*, members of the *Staphylococcus* genus are divided into four groups and 19 species based on cell wall composition and nucleic acid analysis.

The cells of *Staphylococcus* are characterized as spherical (0.5–1.5 μm in diameter), gram-positive, aflagellar, and nonmotile cocci organized as single cells, pairs, tetrads, and clusters. However, they tend to form botryoid clusters. As is the case with other gram-positive bacteria, peptidoglycan and teichoic acid are the two main

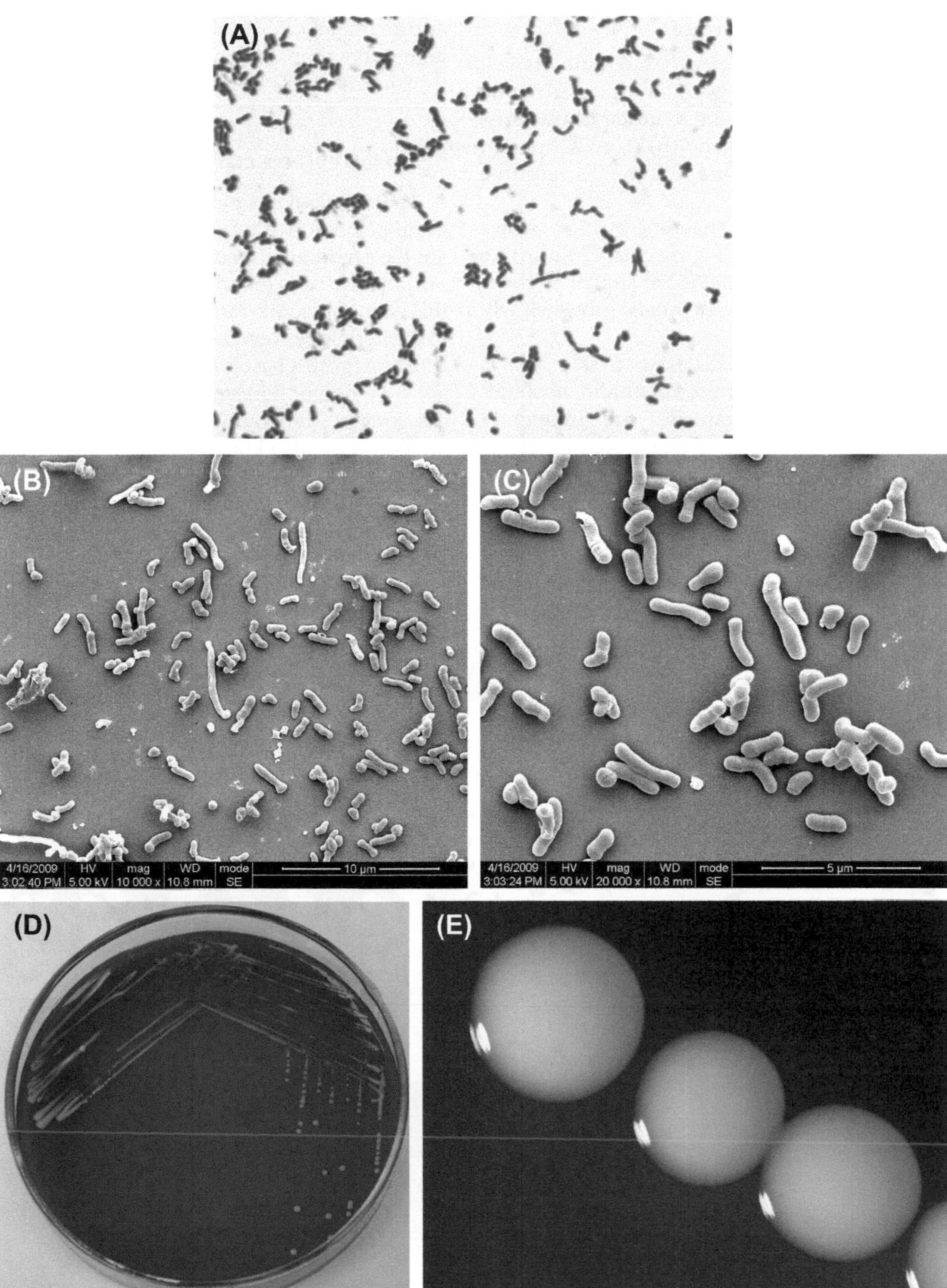

FIGURE 3.10 (A) *Rothia dentocariose* cells (Gram stain). (B) *R. dentocariose* cells (SEM). (C) *R. dentocariose* cells (SEM). The bacteria cells are spherical, similar to *Corynebacterium diphtheriae* in morphology, but can also be filamentous; however, they mostly exist as mixed morphologies. The cell diameter is generally 1.0 μm, but is irregular, as the apical ends of the rod can reach a diameter of 5.0 μm. The culture is almost filamentous on solid media and spherical in broth medium. It is a gram-negative species. (D) *R. dentocariose* colonies (BHI blood agar). (E) *R. dentocariose* colonies (stereomicroscope).

components of the *Staphylococcus* cell wall. The genomic GC content of this genus ranges from 30% to 39%.

Staphylococci are facultative anaerobes, with the exception of *S. saccharolylicus*, which is an anaerobic bacterium. The optimal temperature for the growth of *Staphylococcus* is between 18 °C and 40 °C. Most members of the *Staphylococcus* genus can grow in media containing 10% NaCl. The type species of *Staphylococcus* is *S. aureus*.

3.1.5.1 Staphylococcus epidermidis

S. epidermidis is a gram-positive bacterium. Its cell wall teichoic acid is formed by polymerized glycerol, glucose, and N-acetyl glucosamine. Cellular characteristics are shown in Figure 3.11(A) and (B). The GC content of its genomic DNA ranges from 30% to 37%, and the type strain is ATCC14990.

S. epidermidis is a facultative anaerobe but also grows well under aerobic conditions (Figure 3.11(C) and (D)). Culture conditions for *S. epidermidis* are similar to those of *S. aureus* (see 5.1.1.1), but *S. epidermidis* grows slowly in medium with 10% NaCl.

S. epidermidis mainly colonizes human skin and is a health concern due to its involvement in hospital-acquired infections. The organisms are frequently detected in saliva and dental plaque and are thought to be associated with periodontitis, acute and chronic pulpitis, pericoronitis, dry socket, and angular stomatitis. *S. epidermidis* is sensitive to novobiocin, with the minimum inhibitory concentration at no more than 0.2 mg/L.

Colonies of *S. epidermidis* are round, raised, shiny, gray, and have complete edges. The diameter is approximately 2.5 mm. They usually do not produce a hemolytic zone. Strains that can produce mucus form translucent sticky colonies.

3.1.6 *Streptococcus*

The *Streptococcus* genus makes up the most common gram-positive facultative anaerobic cocci, and its members are the predominant bacteria in the oral cavity. The name *Streptococcus* was given because the bacteria belonging to this genus always arrange themselves into chains. In clinical bacteriology, members of *Streptococcus* are divided into three categories based on their ability to induce hemolysis: α-hemolytic *Streptococcus* (also known as *Streptococcus viridans*), β-hemolytic *Streptococcus*, and γ-hemolytic *Streptococcus* (also known as nonhemolytic *Streptococcus*). In the 2004 edition of *Bergey's Manual of Systematic Bacteriology*, 89 species were attributed to the *Streptococcus* genus.

The most prevalent species in the oral cavity are *Streptococcus salivarius*, *Streptococcus sanguinis*, *Streptococcus mutans*, *Streptococcus sobrinus*, *Streptococcus oralis*, *Streptococcus mitis*, and *Streptococcus gordonii*.

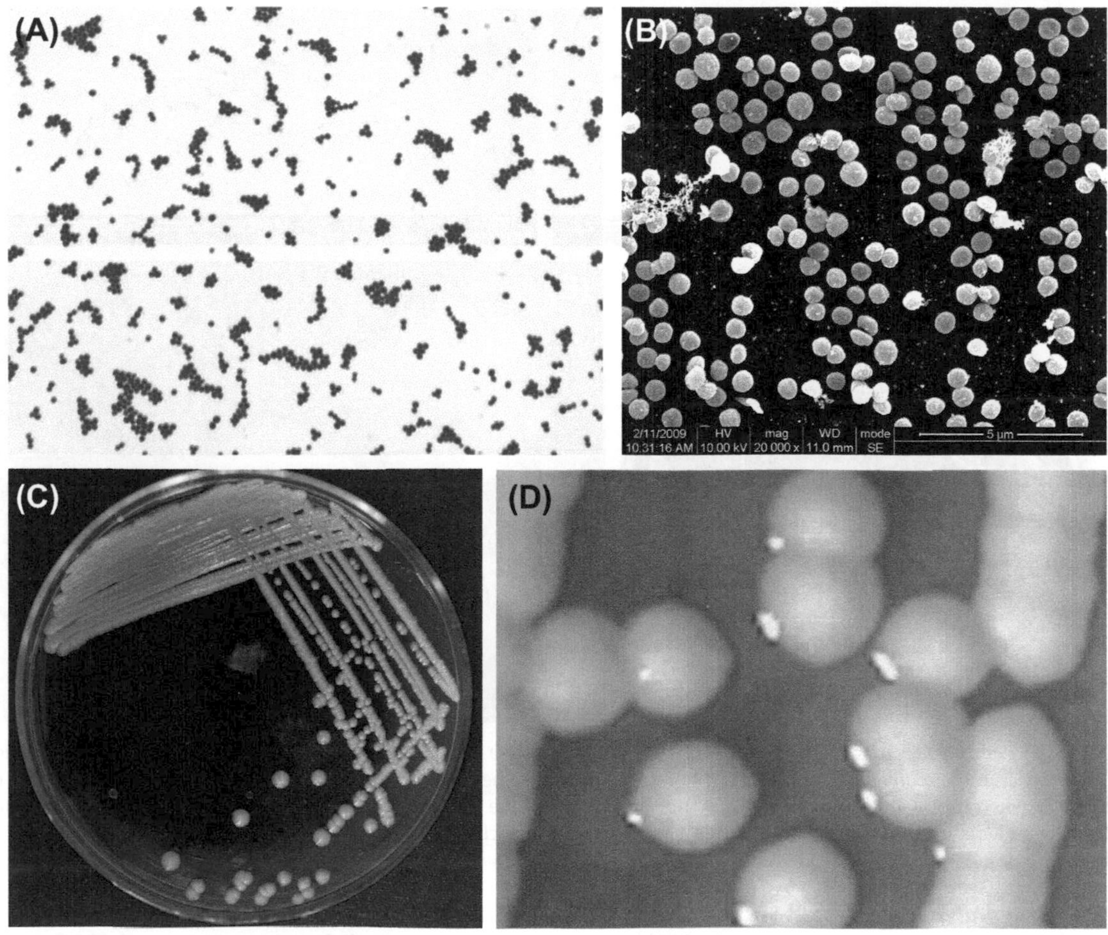

FIGURE 3.11 (A) *Staphylococcus epidermidis* cells (Gram stain). (B) *S. epidermidis* cells (SEM). *S. epidermidis* cells are spherical (0.5–1.5 µm in diameter) and gram-positive. The cocci organize into tetrads and clusters. Single cells are occasionally observed. (C) Colonies of *S. epidermidis* incubated on agar plate. (D) Colonies of *S. epidermidis* (stereomicroscope).

3.1.6.1 Streptococcus salivarius

S. salivarius is a gram-positive coccus (Figure 3.12(A) and (B)). Most strains of *S. salivarius* belong to the Lancefield group K. Their genomic GC content is 39–42%, and the type strain is ATCC7073.

S. salivarius are facultative anaerobes, but the optimal atmosphere condition for bacterial cultures should contain a low percentage of oxygen with 5–10% carbon dioxide. *S. salivarius* grows quickly at 37°C, although it can also grow at 45°C. Cultures require nutrient-rich complex media, such as TS or TPY. The final pH of glucose broth incubated with *S. salivarius* usually falls between pH 4.0 and 4.4. Colony morphology is shown in Figure 3.12(C)–(F).

Biochemical reactions. *S. salivarius* can ferment glucose, sucrose, maltose, raffinose, inulin, salicin, trehalose, and lactic acid. It cannot ferment glycerol, mannitol,

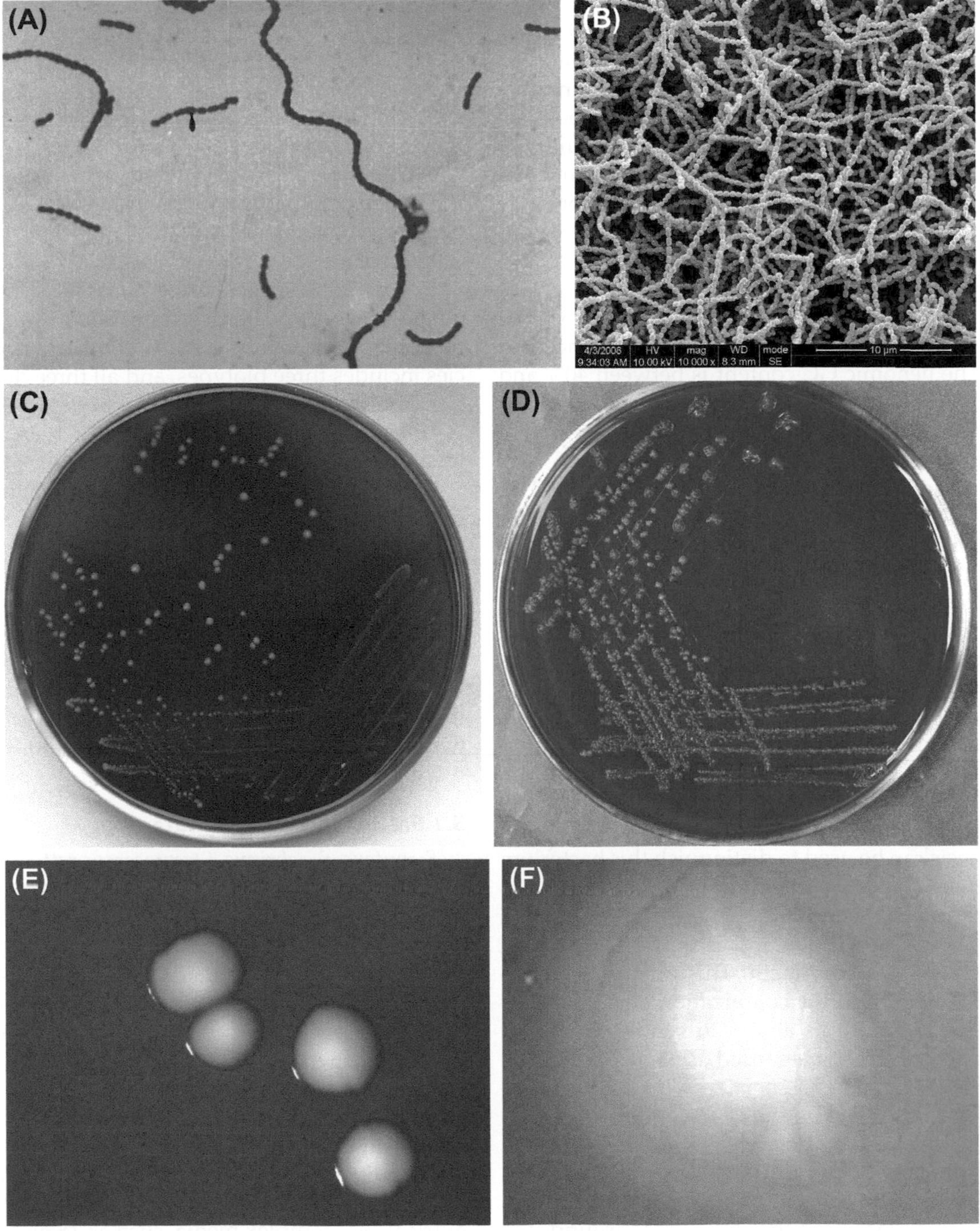

FIGURE 3.12 (A) *Streptococcus salivarius* cells (Gram stain). (B) *S. salivarius* cells (SEM). The cells of *S. salivarius* are spherical or oval (0.8–1.0 μm in diameter) gram-positive cocci organized in short or long chains. (C) α-hemolytic zone of *S. salivarius* incubated on blood BHI agar plate. (D) Rice ball-like colonies of *S. salivarius* incubated on MS agar. (E) Colonies of *S. salivarius* observed under stereomicroscope (incubated on blood BHI agar plate). (F) Colonies of *S. salivarius* isolated from saliva (stereomicroscope).

sorbitol, xylose, and arabinose. Most strains can hydrolyze esculin and urea but not arginine. Meanwhile, most strains can produce acetoin from glucose.

Colonization characteristics. *S. salivarius* is mainly isolated from the oral cavity of humans and animals, and it is part of the normal flora of tongue and saliva microbial communities. Moreover, *S. salivarius* is also detected in fecal and blood samples of endocarditis patients. Research on gnotobiotic animals showed that *S. salivarius* is cariogenic.

The ability to synthesize extracellular polysaccharides determines whether colonies of *S. salivarius* are smooth or rough. On agar plates with sucrose, most strains synthesize soluble fructan and form sticky rice ball-like colonies, a characteristic that can be used to identify *S. salivarius*. Very few strains produce α- or β-hemolytic zones when incubated on agar containing starch or horse blood.

3.1.6.2 Streptococcus sanguinis

S. sanguinis is a gram-positive coccus (Figure 3.13(A) and (B)). Most strains of *S. sanguinis* belong to Lancefield group H. The genomic GC content is between 40% and 46%, and the type strain is ATCC10556 (NCTC 7863).

S. sanguinis is a facultative anaerobe, and the optimal atmospheric condition for cultures should contain 5–10% carbon dioxide. *S. sanguinis* grows quickly at 37 °C, and it cannot grow at 45 °C. Bacterial cultures require nutrient-rich complex media. Blood BHI and TPY media are used for *S. sanguinis* isolation, and MS medium is used for selective culture (Figure 3.13(C)–(E)).

Biochemical reactions. *S. sanguinis* can ferment glucose, maltose, sucrose, trehalose, and salicin to produce acid. Occasionally, it also ferments inulin, raffinose, and sorbitol. *S. sanguinis* does not ferment xylose, arabinose, glycerol, and mannitol. More than 50% of *S. sanguinis* strains hydrolyze esculin. The final pH of glucose broth incubated with *S. sanguinis* is approximately pH 4.6–5.2. Ammonia is produced from arginine hydrolysis, and H_2O_2 synthesis can be used to distinguish this bacterium from *S. mutans*.

Colonization characteristics. *S. sanguinis* is the main component of dental plaque and is only isolated from oral cavities with erupted teeth. This organism is considered to be helpful to the colonization and reproduction of *S. mutan* due to its ability to synthesize PABA. Meanwhile, *S. sanguinis* is regarded as a key probiotic in the oral ecosystem and is associated with healthy periodontal tissues, owing to its ability to synthesize H_2O_2.

S. sanguinis colonies are either smooth or rough, and the diameter of colonies is between 0.7 mm and 1.0 mm. When incubated under aerobic conditions, an α-hemolytic zone can be observed around the colonies of most strains, while a β-hemolytic zone can be observed around the colonies of a few strains. A large number of strains that make up *S. sanguinis* harbor the capacity to produce extracellular polysaccharides above or surrounding their colonies. A liquid-like structure composed of polysaccharides can be observed on these colonies. *S. sanguinis* colonies grown on agar containing a high concentration of sucrose are sticky, hard, and rough. These colonies, with a ground-glass appearance, seem to stretch into the surrounding agar. However, colonies on agar with low sucrose concentration or without sucrose are round, soft, and smooth.

3.1.6.3 Streptococcus gordonii

S. gordonii was classified as *S. sanguinis* serotype II in the past. However, *S. gordonii* lacks the IgA1 protease. It is a gram-positive coccus (Figure 3.14(A)–(C)). The cell wall components are mainly glycerol, teichoic acid, and rhamnose, while its peptidoglycan type is Lys–Ala. The genomic GC content is 40–43%, and the type strain is ATCC10558 (NCTC 7865).

S. gordonii is a facultative anaerobe. Zones of α- or γ-hemolysis can be observed on blood agar, and a green hemolytic zone can be seen on chocolate agar. This species includes three biotypes, and all three do not synthesize catalase. Colonies are shown in Figure 3.14(D)–(F).

Colonization characteristics. *S. gordonii* mainly inhabits the oral cavity and pharynx.

The α-hemolytic zone can be observed surrounding *S. gordonii* colonies incubated on blood agar. Some strains harbor the ability to synthesize extracellular polysaccharides on top of or surrounding their colonies. This layer of polysaccharides appears as a liquid-like structure. In addition, colonies grown on agar containing high sucrose concentration are sticky, hard, and rough. These colonies have a ground-glass appearance and seem to stretch into the surrounding agar. However, colonies grown on agar with low sucrose concentration or without sucrose are round, soft, and smooth.

3.1.6.4 Streptococcus mutans

S. mutans, *S. sobrinus*, *Streptococcus rattus*, *Streptococcus ferus*, *Streptococcus cricetus*, and *Streptococcus macacae* are collectively known as mutans streptococci. These bacteria formerly belonged to serotypes a, b, c, d, e, f, g, or h of *S. mutans*. Their genomic GC content is between 36% and 38%, and the type strain is ATCC25175.

S. mutans is gram-positive, and its colony morphology is shown in Figure 3.15(A)–(E).

S. mutans is a facultative anaerobe, but the optimal atmospheric condition for cultures should be anaerobic or contain only a low percentage of oxygen with 5–10% carbon dioxide. *S. mutans* grows quickly at 37 °C, and some strains can grow at 45 °C. *S. mutans* cultures require nutrient-rich complex media, such as TS and TPY. MSB is used for selective culture (Figure 3.15(F)–(H)). Cells tend to clump or attach to the bottom of the tube when incubated in glucose broth (Figure 3.15(I)), and the final pH of

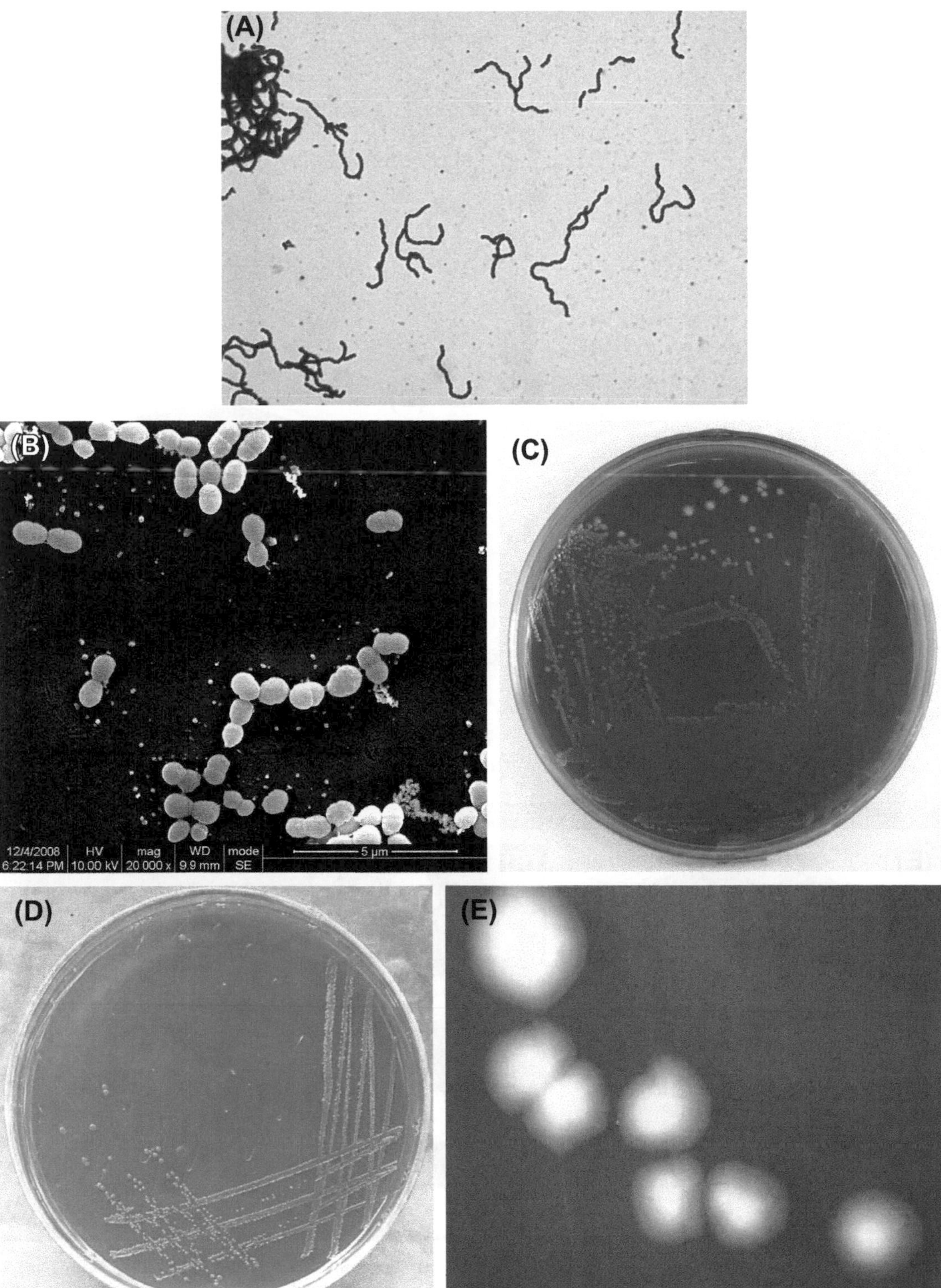

FIGURE 3.13 (A) *S. sanguinis* cells (Gram stain). (B) *S. sanguinis* cells (SEM). *S. sanguinis* cells are spherical or oval (0.8–1.2 μm in diameter), gram-positive cocci that are organized in medium or long chains. Occasionally, the bacteria are rod-shaped or pleomorphic. (C) Colonies of *S. sanguinis* incubated on blood BHI agar plate. (D) Smooth colonies of *S. sanguinis* incubated on MS agar plate. (E) α-hemolytic zone of *S. sanguinis* incubated on blood BHI agar plate (stereomicroscope).

bacterial culture in glucose broth is usually between pH 4.0 and 4.3.

Biochemical reactions. Most strains ferment mannitol, sorbitol, raffinose, lactose, inulin, salicin, mannose, and trehalose to produce acid, but do not ferment arabinose, xylose, glycerol, and melezitose. *S. mutans* hydrolyzes esculin, but not arginine, hippurate, and gelatin. *S. mutans* does not produce H_2O_2.

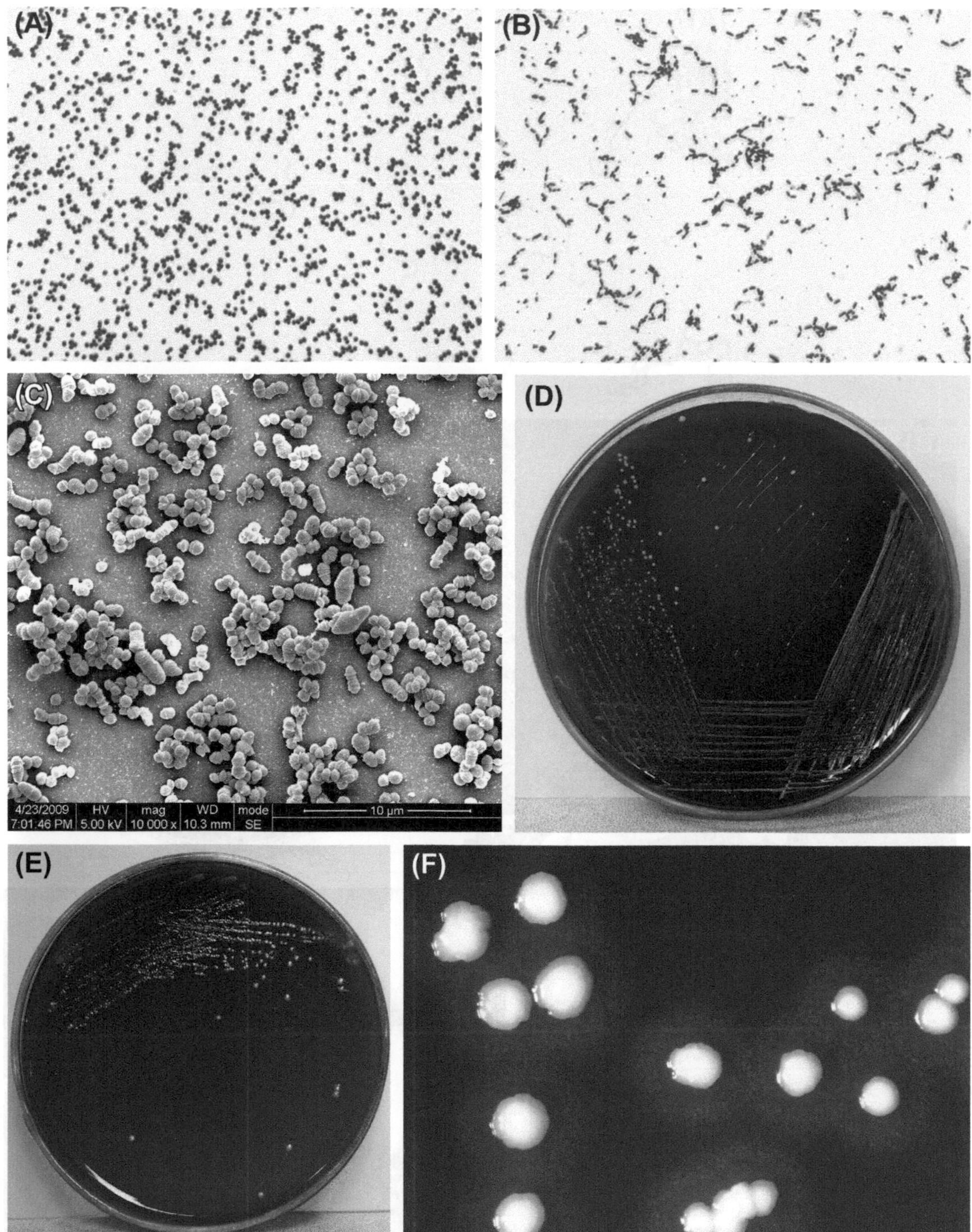

FIGURE 3.14 (A) Spherical cells of *Streptococcus gordonii* (Gram stain). (B) Spear-shaped cells of *S. gordonii* (Gram stain). (C) Proliferating cells of *S. gordonii* (SEM). *S. gordonii* cells are spherical or spear-shaped and organize into chains. The cells are nonmotile and nonsporulating. (D) α-hemolytic zone of *S. gordonii* incubated on blood BHI agar. (E) γ-hemolytic zone of *S. gordonii* incubated on blood BHI agar. (F) Sticky colonies of *S. gordonii* incubated on blood BHI agar (stereomicroscope).

Colonization characteristics. *S. mutans* is mainly isolated from the surface of teeth. It synthesizes a variety of extracellular polysaccharides, including water-soluble and non-water-soluble glucan and fructan from sucrose. These polysaccharides promote bacterial colonization and are key virulence factors in the formation of dental caries. Due to their adhesive capacity, acid production, acid tolerance, and water-soluble glucan production, *S. mutans* has long been regarded as one of the main oral pathogens. It is also involved

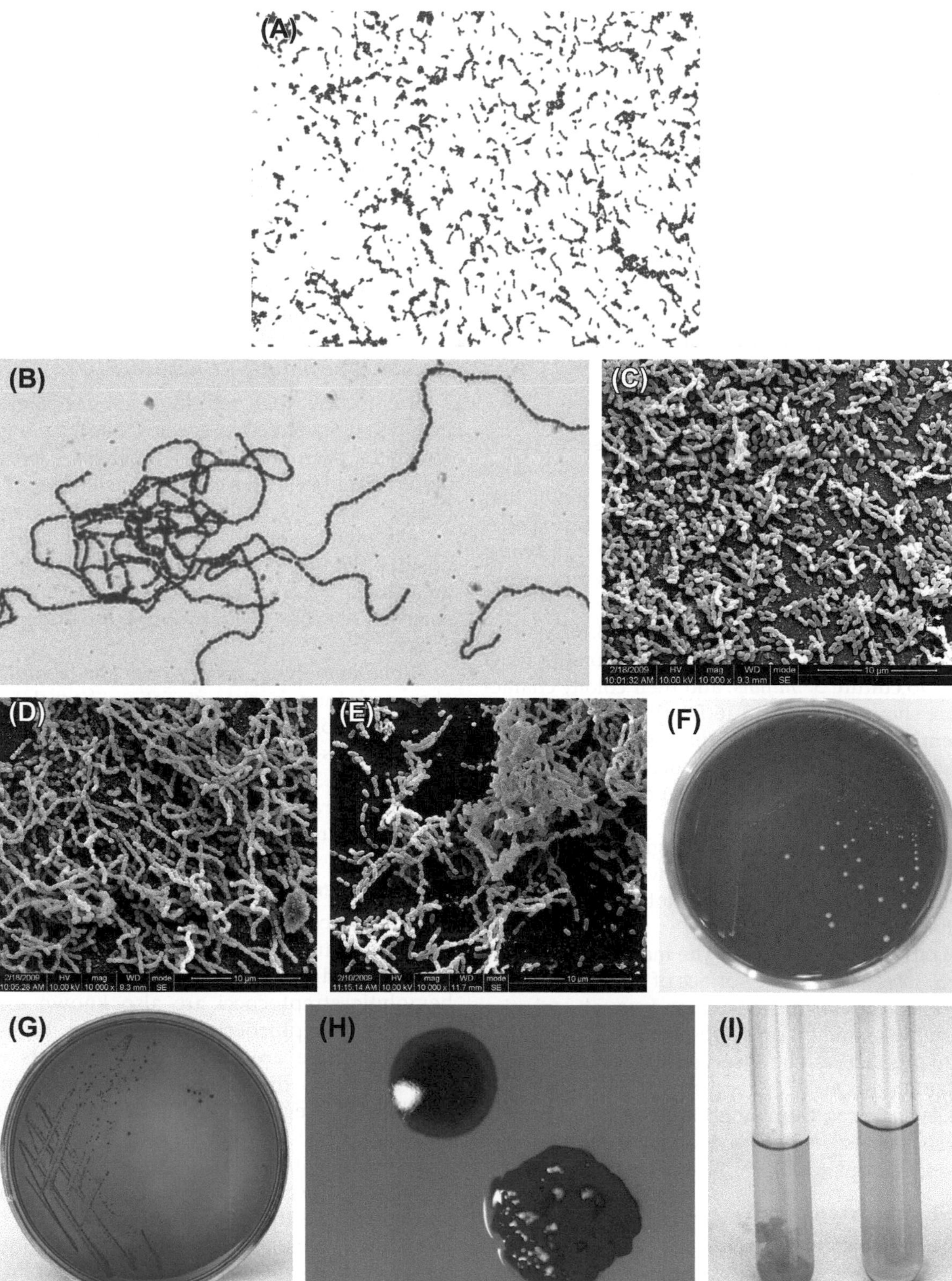

FIGURE 3.15 (A) Spherical cells of *Streptococcus mutans* (Gram stain). (B) Long chain-shaped cells of *S. mutans* (Gram stain). (C) *S. mutans* in short chains (SEM). (D) *S. mutans* in long chains (SEM). (E) Self-curing cells of *S. mutans* (SEM). *S. mutans* are spherical (0.5–0.75 μm in diameter), gram-positive cocci in pairs or chains. Long chains form in broth and short rod-shaped cells (0.5–1.0 μm in length) can be detected in acidic broth and on some solid media. The phenomenon of self-curing bacterial cells can be detected under SEM. (F) Colonies of *S. mutans* incubated on blood BHI agar. (G) Colonies of *S. mutans* incubated on MS agar. (H) Smooth and rough colonies of *S. mutans* incubated on MS agar (stereomicroscope). (I) Bacterial culture of *S. mutans* incubated in TPY broth.

in other secondary infections such as bacteremia and endocarditis.

Colonies of *S. mutans* grown on blood agar after 48 h anaerobic incubation are either regular and smooth or irregular, hard, and sticky. The diameter of colonies is 0.5–1.0 mm. Zones of α- or γ-hemolysis can be observed around colonies of most strains, while β-hemolytic zones can also be observed with the colonies of a few strains. On agar containing sucrose, most strains form stacked colonies (about 1 mm in diameter) with drop-like or myxoid glucan products above or surrounding the colonies. MS is the commonly used selective medium, and both smooth and rough colonies can be observed on the same plate.

3.1.6.5 Streptococcus sobrinus

S. sobrinus was originally classified as serotypes d and g of *S. mutans*, and it is named based on its close relationship with *S. mutans*. Research has shown that *S. sobrinus* has the second highest rate of cariogenicity after *S. mutans*. The GC content of the *S. sobrinus* genome is 44–46% (by Tm method), and its type strain is ATCC33478 (SL).

The cells are gram-positive cocci (Figure 3.16(A)–(C)). Cultures of *S. sobrinus* grow under similar conditions as those used to culture *S. mutans*, and their colony characteristics are shown in Figure 3.16(D)–(F).

Biochemical reactions. *S. sobrinus* is able to ferment mannitol, inulin, and lactose to produce acid, but its ability to ferment sorbitol, D-melibiose, and raffinose varies by strain. Moreover, most strains of *S. sobrinus* can produce H_2O_2 but cannot metabolize arginine to produce ammonia. Most strains also cannot hydrolyze aesculin and do not synthesize obvious amounts of extracellular polysaccharide.

Colonization characteristics. The main site of colonization of *S. sobrinus* is on the surface of human teeth.

S. sobrinus can form stacked and rough colonies (about 1 mm in diameter) on sucrose agar plates. Liquid-like glucan products can be observed above or surrounding these colonies. On TPY agar plates or BHI blood agar plates, *S. sobrinus* can form smooth sticky colonies, and α-hemolysis can also be detected for some strains growing on blood agar plates.

3.1.6.6 Streptococcus oralis

The cells of *S. oralis* are gram-positive (Figure 3.17(A)), spherical, and arranged in short chains. In addition, *S. oralis* cells are nonmotile, have no capsule, and do not form spores. The genomic GC content of *S. oralis* is 40% (Tm method), and its type strain is NCTC11427 (LUG1, PB182).

S. oralis is a facultative anaerobe and most commonly grown on TPY agar (Figure 3.17(B)). Cells of this species can grow in medium containing 0.0004% crystal violet, and α-hemolytic reaction can be detected when colonies are grown on blood agar plates (Figure 3.17(C)). Characteristic *S. oralis* colonies are shown in Figure 3.17(D).

S. oralis can reduce tetrathionate but does not produce catalases. In fact, *S. oralis* is relatively biochemically inactive. This species is mainly isolated from the human oral cavity and is a common member of the oral microflora.

Streptococci are clinically divided into three major categories: α-hemolytic, β-hemolytic, and γ-hemolytic.

3.1.6.7 α-hemolytic Streptococcus

Streptococci that fall into the α-hemolytic group include all *Streptococcus* species that form a grass-green hemolytic zone around their colonies when grown on blood agar plates. This category includes species such as *S. sanguis*, *S. mitis*, and *Streptococcus vestibularis*.

Like other streptococci, α-hemolytic cells are gram-positive (Figure 3.18(A)), spherical, and nonmotile. The great majority of cells are organized in pairs or short chains. Cells observed by SEM are shown in Figure 3.18(B).

Alpha-hemolytic streptococci are facultative anaerobes and form a characteristic α-hemolytic zone on blood agar plates. The α-hemolytic zone appears as a narrow grass-green zone that is observed around colonies (Figure 3.18(C)). Members of this group are also known as grass-green streptococci. Colonies observed by stereomicroscope are shown in Figure 3.18(D).

3.1.6.8 β-hemolytic Streptococcus

Streptococcal species that fall into the β-hemolytic category include all species that can form a β-hemolytic zone, including *S. pyogenes* and *S. agalactiae*. Beta-hemolytic streptococci are also known as pyogenic hemolytic streptococci and have the highest pathogenicity. These are the causative agents of various oral infectious diseases, including phlegmon in the maxillofacial region, acute tonsillitis, and periodontal abscess.

Like other *Streptococcus* species, cells of β-hemolytic streptococci are gram-positive (Figure 3.19(A)), rounded, nonmotile, and the great majority organize themselves into short chains. However, most cells in liquid culture form long chains. Cells observed by SEM are shown in Figure 3.19(B).

Beta-hemolytic streptococci are facultative anaerobes that form a broad and completely transparent hemolytic zone around colonies grown on blood agar plates (Figure 3.19(C)). Colonies observed by stereomicroscope are shown in Figure 3.19(D).

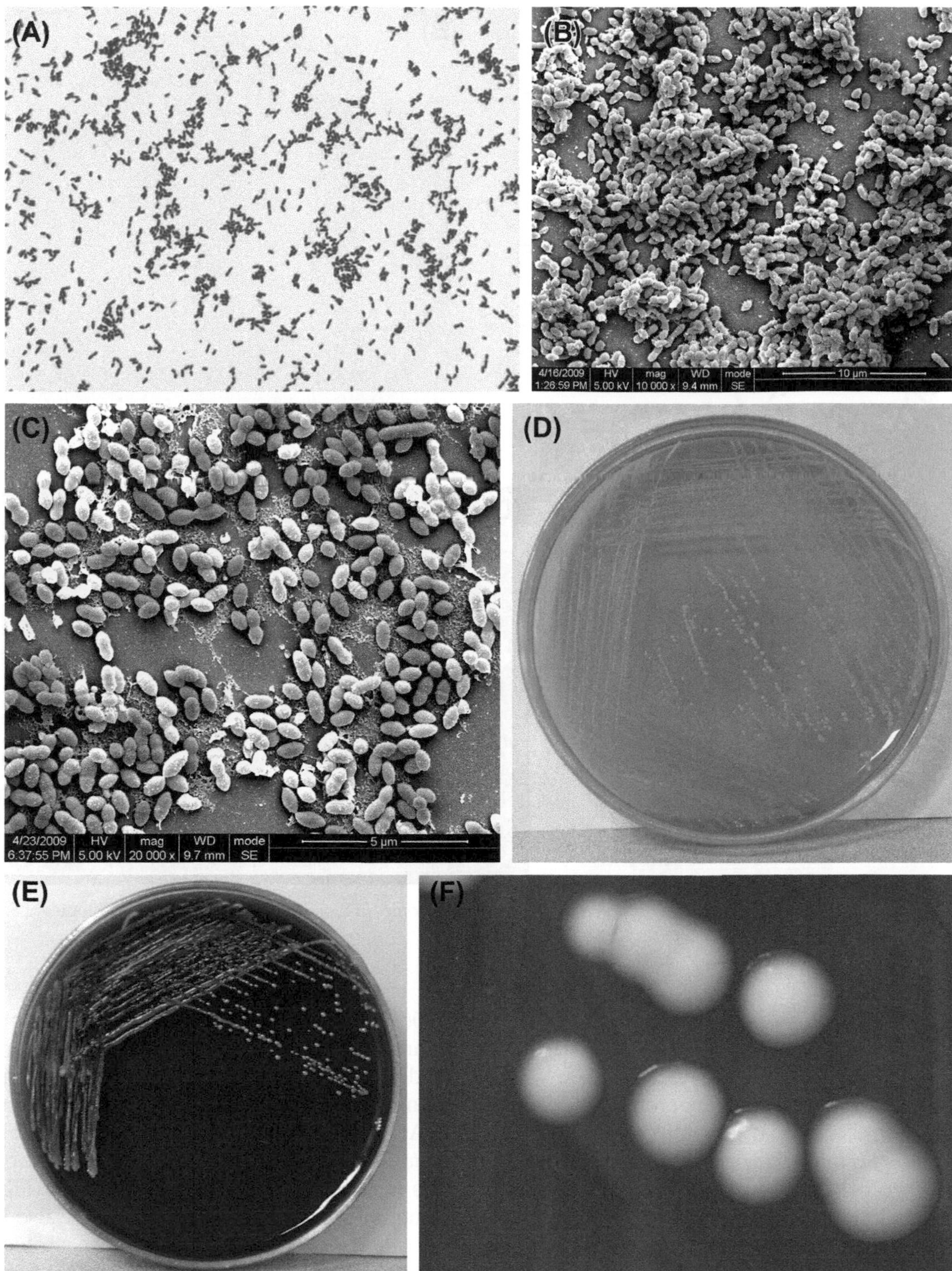

FIGURE 3.16 (A) *Streptococcus sobrinus* cells (Gram stain). (B) *S. sobrinus* cells (SEM). The cells of *S. sobrinus* are gram-positive, spherical (about 0.5 µm in diameter), arranged in pairs or in chains, and often form long chains. (C) *S. sobrinus* cells (SEM). (D) Colonies of *S. sobrinus* (TPY agar plate). (E) Colonies of *S. sobrinus* (BHI blood agar plate). (F) Alpha-hemolytic colonies of *S. sobrinus* (BHI blood agar plate, stereomicroscope).

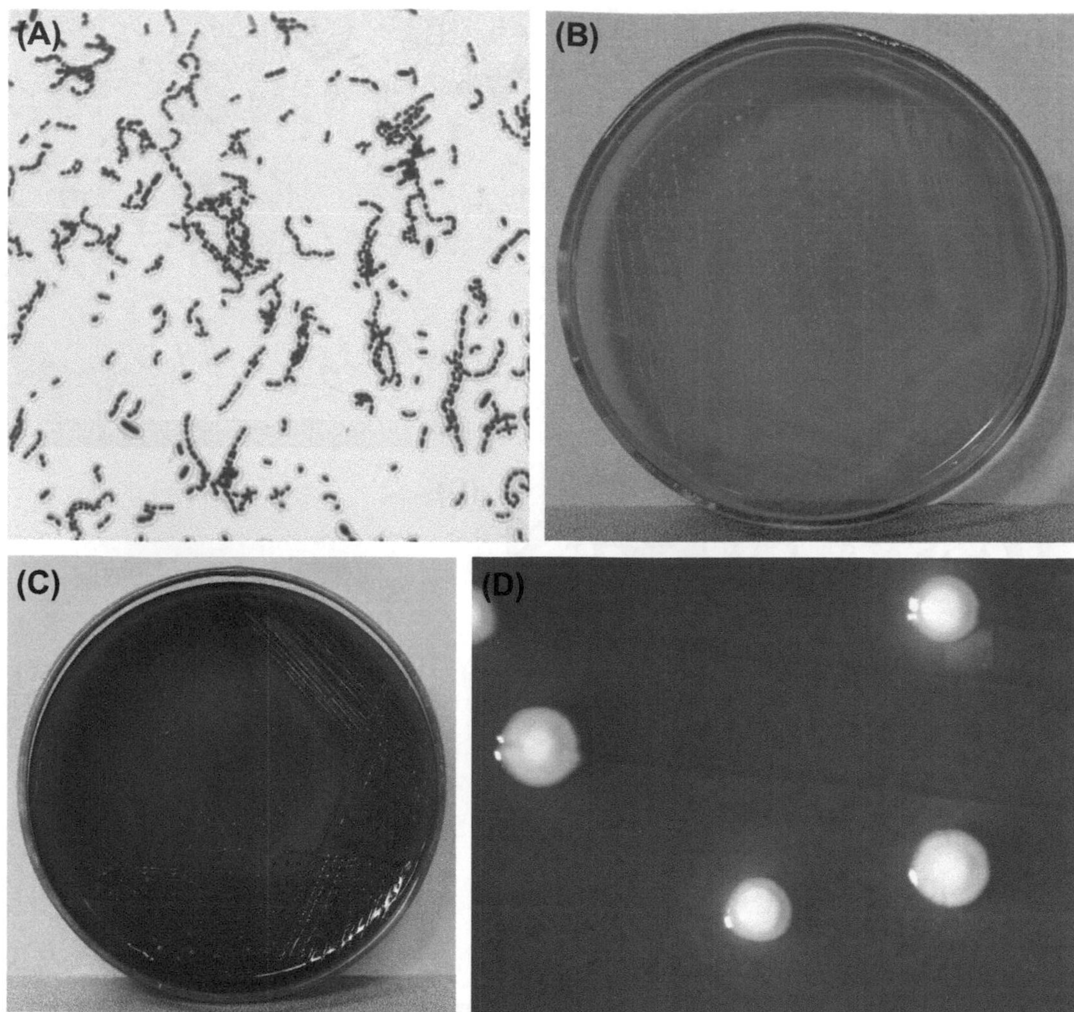

FIGURE 3.17 (A) Cells of *Streptococcus oralis* (gram-positive coccus). (B) Colonies of *S. oralis* (TPY agar plate). (C) Colonies of *S. oralis* (BHI agar plate). (D) Colonies of *S. oralis* (stereomicroscope).

3.2 GRAM-NEGATIVE BACTERIA

3.2.1 *Leptotrichia*

Leptotrichia is a gram-negative anaerobic bacillus and is a very commonly observed genus in the human oral cavity. The genus *Leptotrichia* was first found in 1896 and was named leptothrix as it was isolated from the rabbit uterus. For a long time, *Leptotrichia* were considered opportunistic pathogens until recent reports that indicated that they may be pathogenic. *Leptotrichia* can be isolated from the oral cavity and are mainly found in bacterial biofilms. It can also separate from the vagina and the uterus of pregnant women.

The *Leptotrichia* cell measures 0.8–1.5 μm × 5–20 μm. They can be straight or curved rod shapes. The ends of the cell (either one or both ends) can be sharp or rounded. Cells normally organize as pairs or in a chain (Figure 3.20(A)–(C)). The cells do not produce spores and are nonmotile. Fresh cultures can be stained gram-positive. Under the light microscope, both gram-negative and gram-positive cells can be observed on a single slide.

After culturing in anaerobic blood agar for 1–2 days, *Leptotrichia* can form 1–2 mm, raised, and transparent colonies with smooth and filamentous edges (Figure 3.20(D)). Sometimes polymorphous colonies are also formed.

Leptotrichia grow best under anaerobic conditions. Cultures require 5–10% CO_2. The ideal temperature for culture growth is between 35 °C and 37 °C, while *Leptotrichia* cells stop growing when temperatures drop below 25 °C. The ideal pH for culturing these cells is between pH 7.0 and 7.4. Growth is not inhibited by 20% bile.

Leptotrichia is biochemically active. It can ferment amygdalin, cellobiose, fructose, glucose, maltose, mannose, melezitose, salicin, sucrose, and trehalose to produce acid.

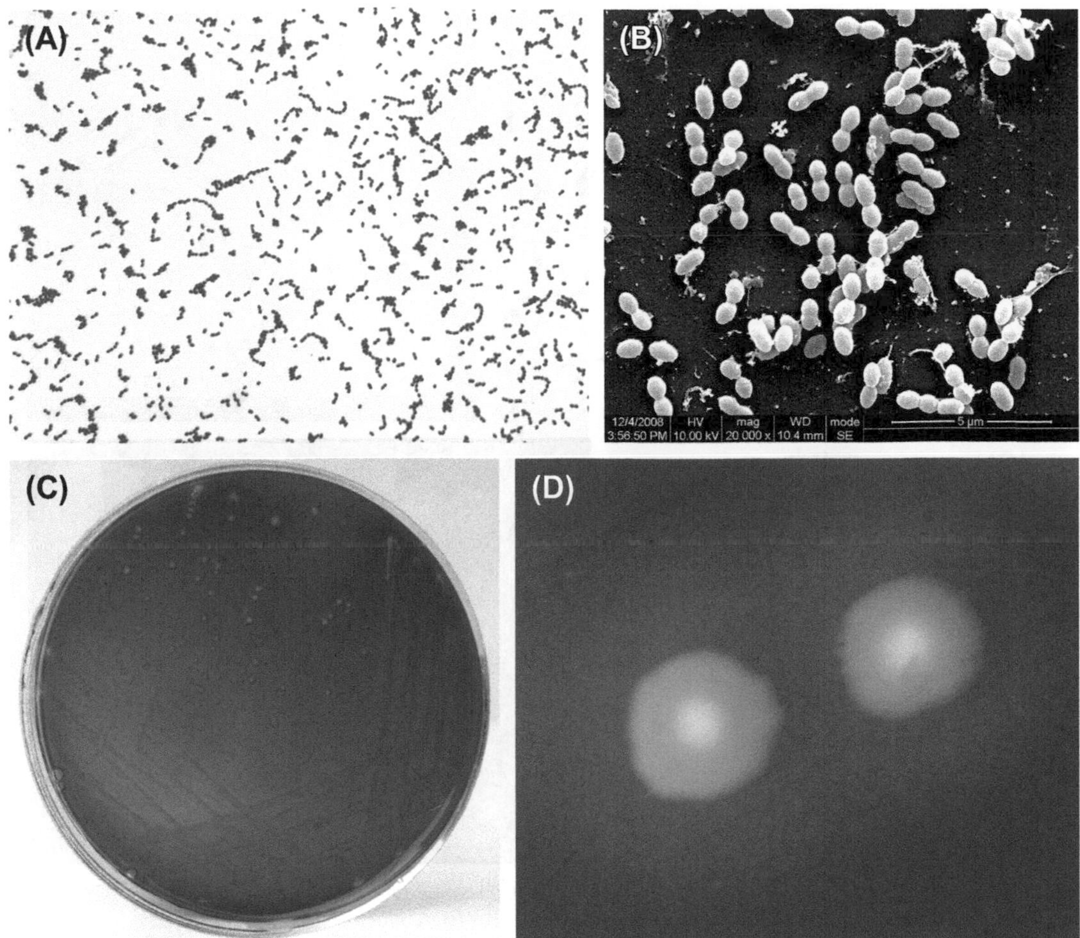

FIGURE 3.18 (A) Cells of α-hemolytic streptococcus (Gram stain). (B) Cells of α-hemolytic streptococcus (SEM). (C) Colonies of α-hemolytic streptococcus (BHI blood agar plate). (D) Colonies of α-hemolytic streptococcus (stereomicroscope).

The terminal products of lactose and starch fermentation are variable. *Leptotrichia* does not ferment arabinose, dulcitol, glycerol, inositol, inulin, mannitol, melibiose, raffinose, rhamnose, ribose, sorbitol, and xylose. The cells do not produce indole, catalase, urease, H_2S, phospholipase, and ammonia gas.

The percent GC in *Leptotrichia* DNA is 25% (by Tm or Bd). The type strain is ATCC14201.

3.2.2 *Veillonella*

Veillonella are gram-negative anaerobic cocci and belong to the family *Veillonellaceae*. Strains detected in oral cavities include *Veillonella parvula*, *Veillonella atypic*, and *Veillonella dispar*.

3.2.2.1 Veillonella parvula *subsp.* parvula

V. parvula subsp. *parvula* are gram-negative anaerobic cocci and are among the most common bacteria in the oral cavity. Cell characteristics are shown in Figure 3.21(A) and (B). The DNA GC content is 38% when analyzed by Tm or 41% when analyzed by Bd. The type strain is ATCC10790.

V. parvula subsp. *parvula* is a strict anaerobe. Several strains require putrescine and cadaverine in their growth medium. Characteristic colonies are shown in Figure 3.21(C) and (D).

The cells are relatively biochemically inactive when tested using classical biochemical tests. They are unable to ferment carbohydrate to acid and do not produce indole. They appear negative with the catalase test but are able to reduce nitrate to nitrite.

V. parvula subsp. *parvula* is detected in saliva, on the tongue, and in plaques. They are able to utilize lactate produced by *Streptococcus mutans*, and are thus considered as beneficial bacteria in dental plaques. They form part of the normal human gut flora.

V. parvula subsp. *parvula* require strictly anaerobic conditions, forming small gray-white colonies on the surface of BHI blood agar.

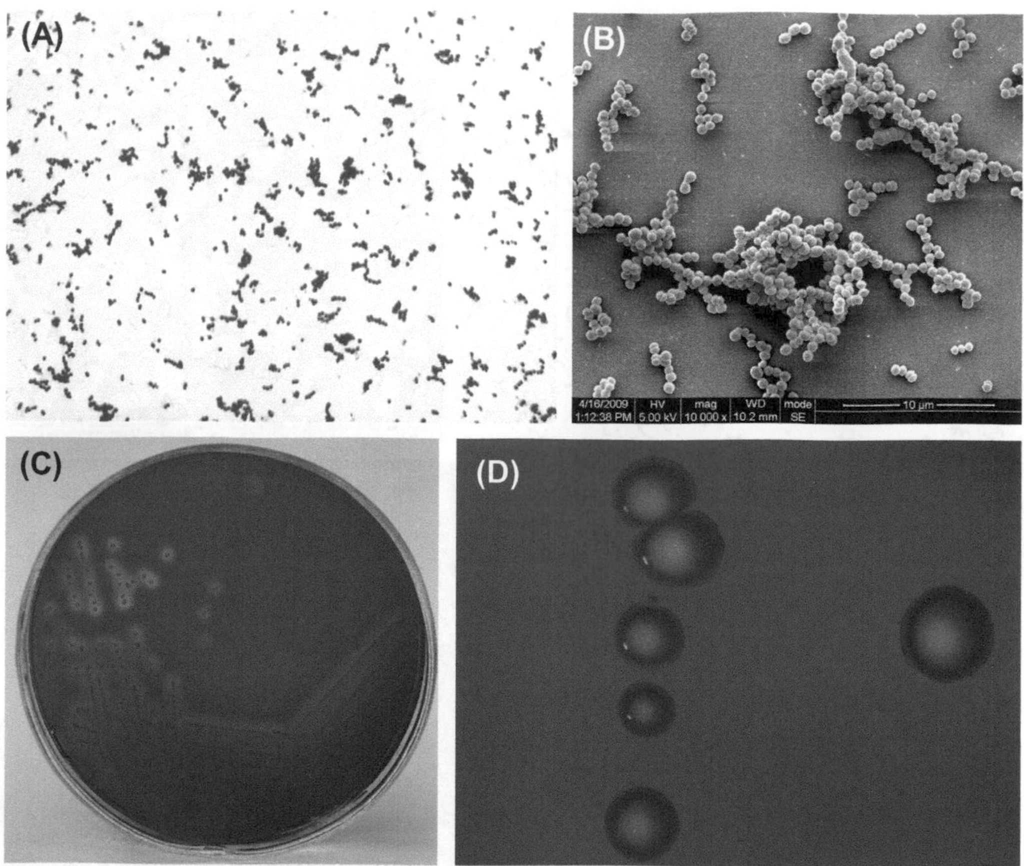

FIGURE 3.19 (A) Cells of β-hemolytic streptococcus (Gram stain). (B) Cells of β-hemolytic streptococcus (SEM). (C) Colonies of β-hemolytic streptococcus (BHI blood agar plate). (D) Colonies of β-hemolytic streptococcus (stereomicroscope).

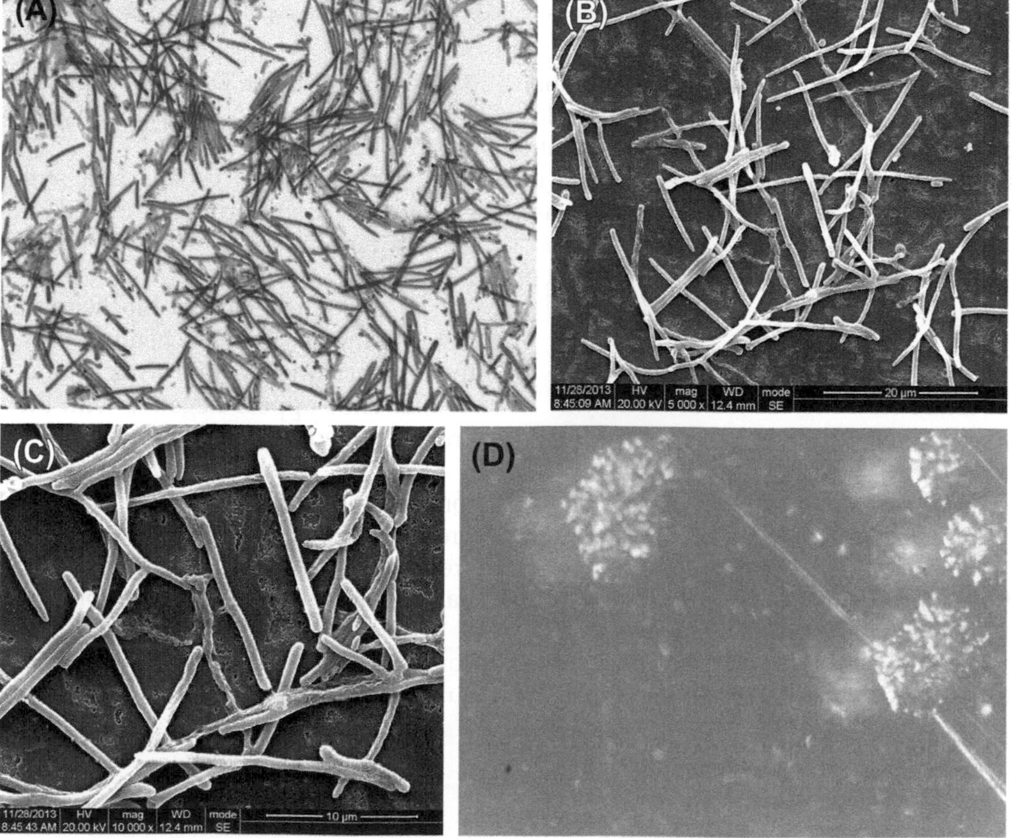

FIGURE 3.20 (A) *Leptotrichia* cells (Gram stain). (B) *Leptotrichia* cells (SEM). (C) *Leptotrichia* cells (SEM). (D) *Leptotrichia* colonies.

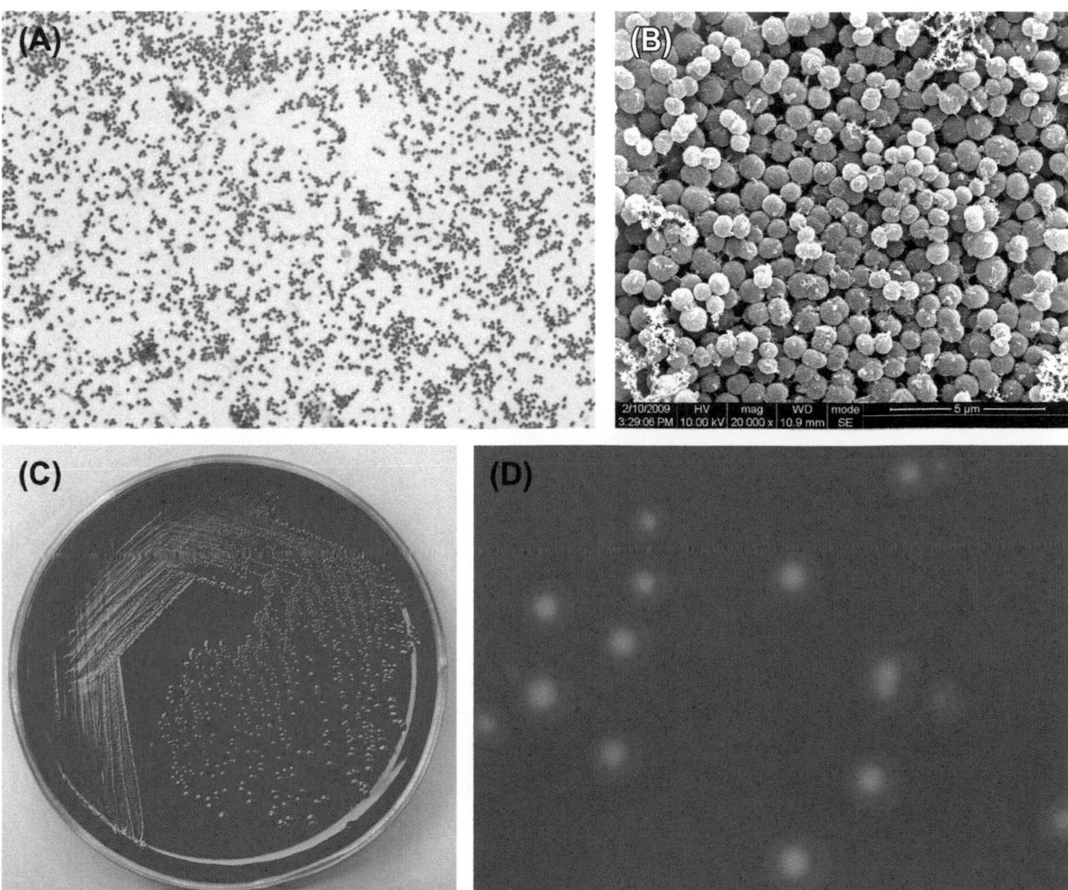

FIGURE 3.21 (A) *Veillonella parvula* subsp. *parvula* cells (Gram stain). (B) *V. parvula* subsp. *parvula* cells (SEM). *V. parvula* subsp. *parvula* cells are spherical and often arranged in piles or clumps. The cells are gram-negative, but can show up as gram-positive in immature cultures. (C) *V. parvula* subsp. *parvula* colonies (BHI blood agar). (D) *V. parvula* subsp. *parvula* colonies (stereomicroscope).

CHAPTER

4

Subgingival Microbes

4.1 GRAM-POSITIVE BACTERIA

4.1.1 *Enterococcus*

Enterococci are facultative gram-positive cocci and belong to Lancefield group D. *Enterococci faecalis* is also called *Streptococcus faecalis* and is the most common species in the genus *Enterococcus*. In recent years, it has been closely studied due to its high detection rate in infected root canals.

4.1.1.1 Enterococcus faecalis

E. faecalis cells are gram-positive, oval (0.5–1 μm in diameter), and nonmotile. Most cells are arranged in pairs or as short chains (Figure 4.1(A)). Cellular morphology by SEM is shown in Figure 4.1(B). The GC content of its DNA is 33.5%. The type strain is NCTC775 (ATCC19433, NCDO5681).

E. faecalis is a facultative anaerobe. Cells of this species can form smooth, nontransparent, white or creamy, spherical colonies on common nutrient agar plates (Figure 4.1(C)). However, colonies formed on PSE agar plates containing cholate, esculin, and sodium azide (the selective agar medium for Pfizer *Enterococci*) are brownish black with brown aureole (Figure 4.1(D)). This can be used as a characteristic to distinguish *E. faecalis* from other bacteria. Colonies observed by stereomicroscope are shown in Figure 4.1(E).

E. faecalis can ferment most carbohydrates. The main acid produced by glucose fermentation is lactic acid. *E. faecalis* can also hydrolyze arginine to produce ammonia.

4.1.2 *Eubacterium*

Eubacterium is a genus of gram-positive nonsporulating strictly anaerobic bacilli. The name of the genus is still disputed. *Bergey's Manual of Systematic Bacteriology* vol. 2 (1986) points out that the Greek prefix *eu* means good, useful, rather than true. Therefore, the author believes that *Eubacterium* is the more appropriate name. Currently, bacteria species detected in the oral cavity that belong to this genus include *Eubacterium alactolyticum, Eubacterium saburreum, Eubacterium lentum, Eubacterium limosum, Eubacterium nodatum, Eubacterium brachy, Eubacterium timidum, Eubacterium saphenus*, and *Eubacterium minutum*.

Cells can be homogeneous or polymorphous rod-shaped. No spores are produced. Cells are gram-positive, but Gram staining old cultures and cultures that have produced acid in the culture medium will yield negative results.

Eubacteria are strictly anaerobic. Culturing cells can be difficult due to this bacteria's strict anaerobic demands, and some strains can only grow in prereduced medium. The optimum growth temperature is 37°C, while the optimum pH is 7.0.

Eubacteria are chemoheterotrophs and produce energy from mixed organic acids produced by carbohydrates or protein metabolism. These mainly consist of butyric acid, acetic acid, and formic acid.

Most Eubacteria from the oral cavity are relatively biochemically inactive. In most cases, cells test negative for catalase and do not hydrolyze hippurate. Carbohydrate fermentation, indole production, nitrate reduction, esculin hydrolysis, and other biochemical tests can help differentiate the different species in the genus.

Eubacteria mainly colonize the saliva and plaque as a member of the normal oral microflora. *Eubacterium lentum* and *E. limosum* can be detected in the oral cavity. *Eubacterium nodatum, E. brachy, E. timidum, E. saphenus*, and *E. minutum* are new species isolated from subgingival plaque of patients with periodontitis and are considered as potential periodontal pathogens. The GC content in *Eubacterium* DNA is 30–55% (analyzed by Tm), and the percentage in the type species is 47% (by Tm).

4.1.2.1 Eubacterium lentum

In 1999, Kageyama et al. called for a change in the classification of *E. lentum* and proposed that it

Atlas of Oral Microbiology
http://dx.doi.org/10.1016/B978-0-12-802234-4.00004-5

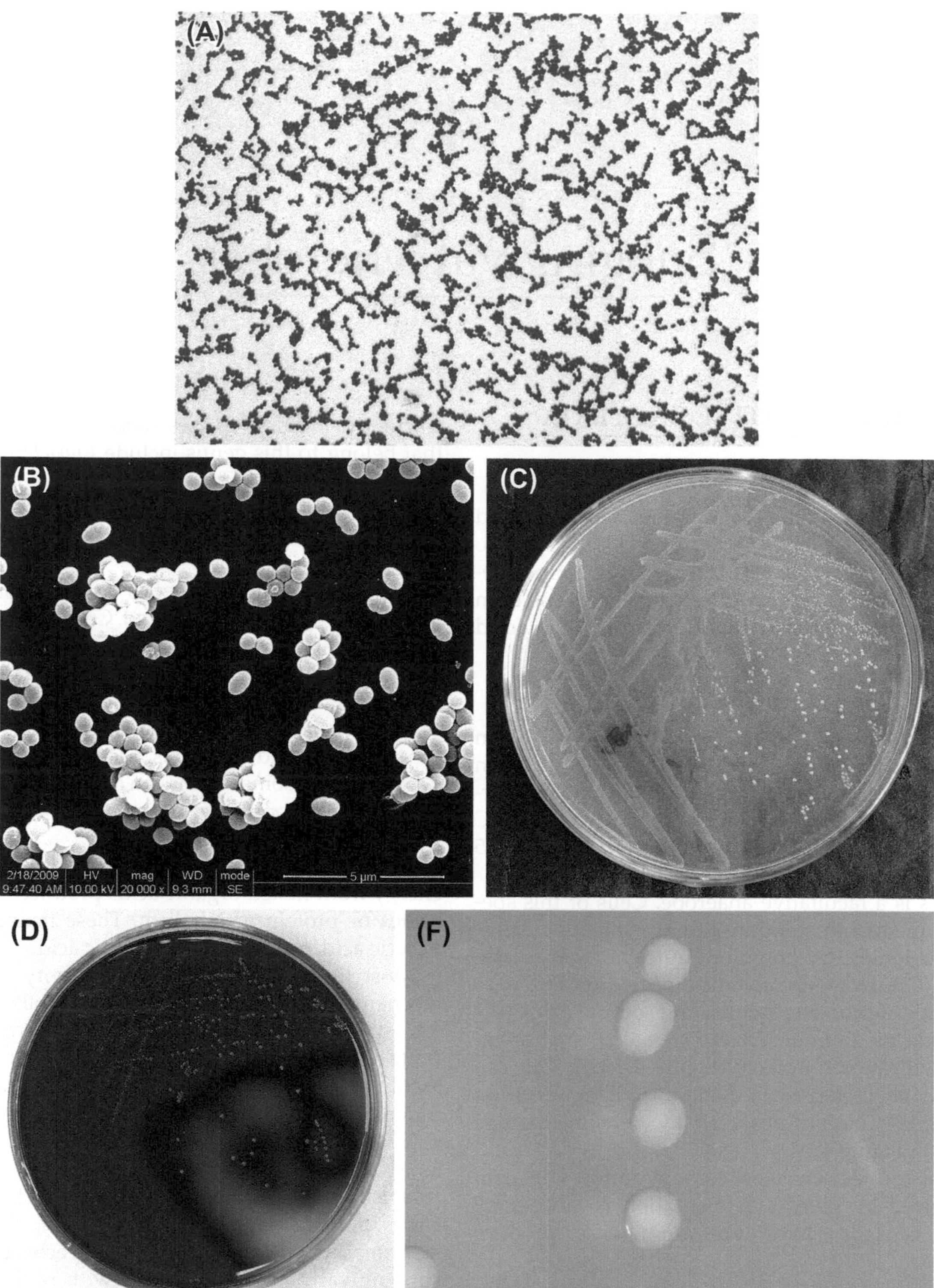

FIGURE 4.1 (A) Cells of *E. faecalis* (gram-positive cocci). (B) Cells of *E. faecalis* (SEM). (C) Colonies of *E. faecalis* (common nutrient agar plate). (D) Colonies of *E. faecalis* (PSE agar plate). (E) Colonies of *E. faecalis* (stereomicroscope).

be grouped with *Eggerthella lenta*, the type species of genus *Eggerthella*.

E. lentum is a gram-positive, irregular, nonsporulating, strictly anaerobic bacillus (Figure 4.2(A) and (B)). As a strict anaerobe, most strains can grow between 30 °C and 45 °C, while some strains can grow at 25 °C. Arginine can promote bacterial growth. Characteristic colonies are shown in Figure 4.2(C) and (D).

E. lentum does not ferment carbohydrates. The cells do not hydrolyze esculin, hippurate, and gelatin.

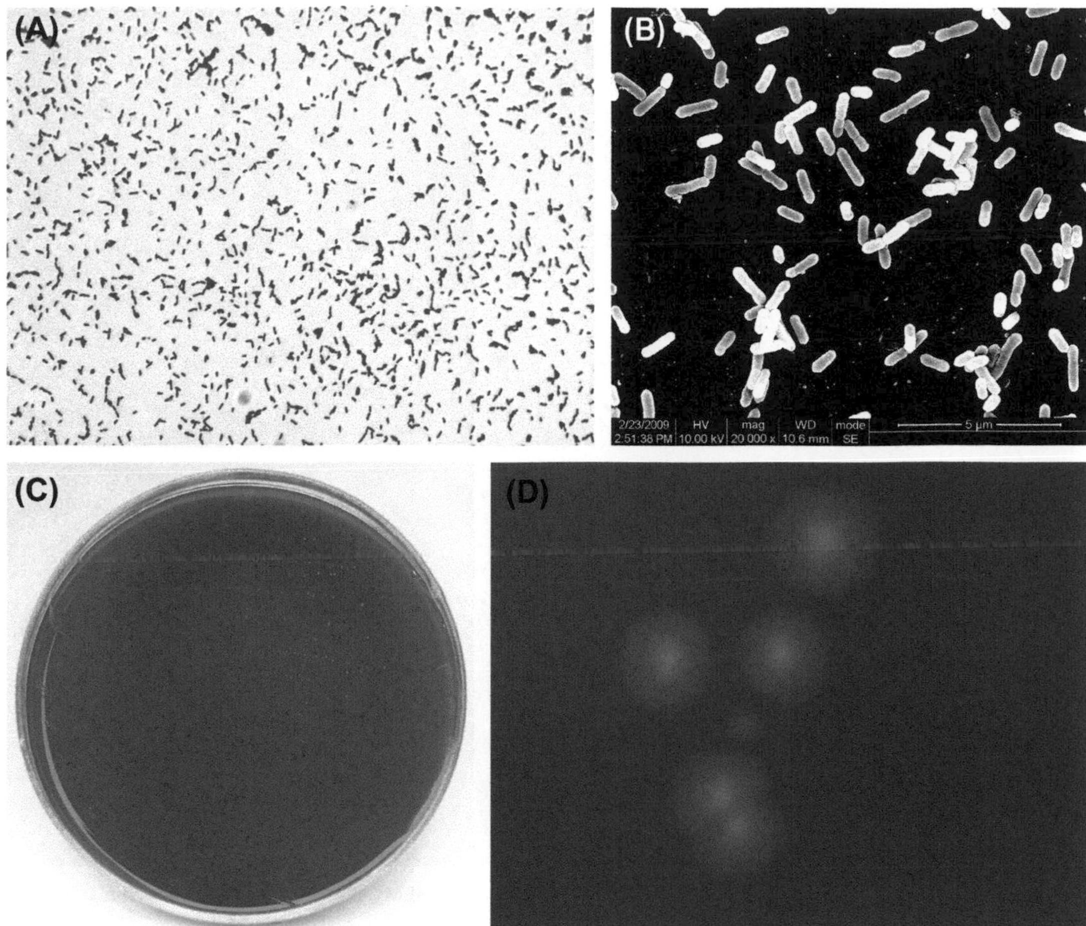

FIGURE 4.2 (A) *Eubacterium lentum* cells (Gram stain). (B) *E. lentum* cells (SEM). (C) *E. lentum* colonies (BHI blood agar). (D) *E. lentum* colonies (stereomicroscope). *E. lentum* on horse blood agar forms surface colonies that are 0.5–2 mm in diameter, rounded, convex or convex, low, dim, and dark or lustrous, translucent or opaque, smooth, wedge-shaped, or neat-edged. A striped appearance can be observed under incident light.

Ammonia can be produced from arginine, and H_2O_2 can be produced from agar medium containing 1% arginine. Under anaerobic conditions, the bacteria can produce H_2S in the beveled bottom of triple sugar iron agar but cannot produce H_2S in SIM culture medium. The GC is 62% in DNA, and the type strain is JCM 9979 (= DSM2243 = ATCC25559 = NCTC11813).

4.1.3 *Peptostreptococcus*

Peptostreptococcus is the most common gram-positive anaerobic coccus in the human oral cavity and in the clinic. *Peptostreptococcus anaerobius* and *Peptostreptococcus micros* are the most commonly encountered species in this genus.

4.1.3.1 Peptostreptococcus anaerobius

The cells of *P. anaerobius* are gram-positive (Figure 4.3(A)), spherical, approximately 0.5–0.6 μm in diameter, and arranged in pairs or chains. Cells in early cultures have been observed to form long chains. Cells observed by SEM are shown in Figure 4.3(B). The GC content of the *P. anaerobius* genome is 33–34%, and its type strain is ATCC27337.

The optimal temperature for *P. anaerobius* growth is 37 °C, and cells of this species do not grow well at 25 °C or 30 °C, and do not grow at all at 45 °C. Growth is stimulated by 0.02% polysorbate-80. *P. anaerobius* cells form pin-point-like or rounded (about 1 mm in diameter), raised, white, glossy, nontransparent colonies with a smooth surface, without hemolytic zone (Figure 4.3(C)). Colonies formed on the surface of BHI supplemental medium without addition of blood are gray. Broth cultures of *P. anaerobius* are usually not muddy, and granulated or viscous precipitates can be observed. Colonies observed by stereomicroscope are shown in Figure 4.3(D).

P. anaerobius is relatively biochemically inactive, and cells usually do not ferment carbohydrates. The main acidic metabolic end products in PVG liquid culture of the type strain are acetic acid, isobutyric acid, butyric acid, isovaleric acid, and isocarproic acid. *P. anaerobius* can produce CO_2 and H_2 from pyruvates under anaerobic conditions. In deep glucose agar, *P. anaerobius* can produce a large amount of gas and generate ammonia from peptone.

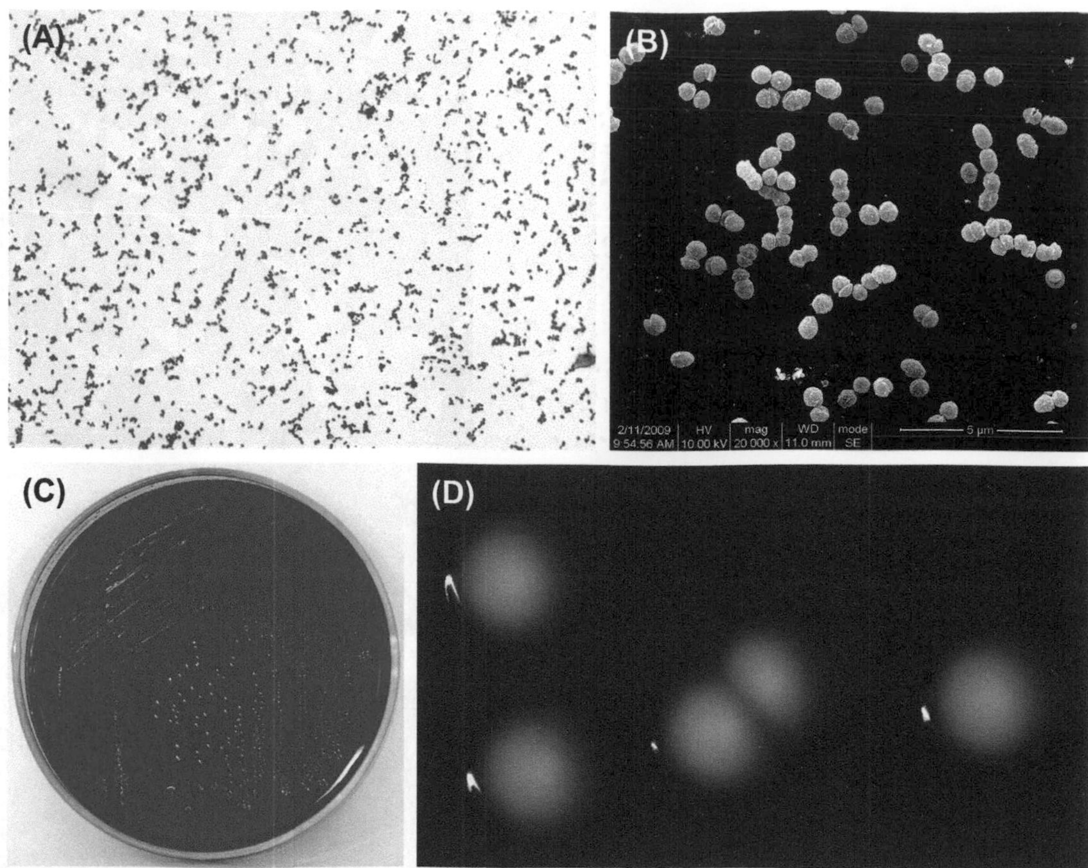

FIGURE 4.3 (A) Cells of *P. anaerobius* (Gram stain). (B) Cells of *P. anaerobius* (SEM). (C) Colonies of *P. anaerobius* (BHI blood agar plate). (D) Colonies of *P. anaerobius* (stereomicroscope).

Dental plaque and gingival sulcus are the main habitats for *P. anaerobius* in the oral cavity. Moreover, *P. anaerobius* is often detected in clinical samples of periodontitis, pulpitis, and pericoronitis.

4.1.4 *Propionibacterium*

Propionibacterium is a genus of gram-positive polymorphic bacilli that do not sporulate. The genus can be divided into two groups: one that lives on the skin, including inside the oral and intestinal tracts, named the sores and blisters group or *P. acnes*; and another that lives in dairy products, cheese or green fodder named dairy group or typical propionibacteria.

The cells are polymorphic in form, many have two rounded ends, and others are shaped like *Corynebacterium diphtheria*, with one rounded end and one tapered end. Cells are 0.5–0.8 μm in diameter, 1–5 μm in length, and form two branches or branched rods. Cocci are observed in old cultures, arranged as single cells, in pairs, in chains, or in "Y"- or "V"-shaped branched chains. Cells are gram-positive, but some cells can also stain gram-negative.

Members of this genus are anaerobic or microaerophilic bacteria. The highest rate of growth takes place 48 h after inoculation. The Propionibacteria are chemoheterotrophic bacteria that need a complex nutritional medium such as BHI agar in order to be cultured. Most species can grow in dextrose broth containing 20% bile or 6.5% NaCl. *Propionibacterium acnes* colonies on the surface of agar can produce colorful pigmentation including white, gray, pink, red, or yellow.

The main acid metabolites are propionic acid and acetic acid when cultured in PYG broth. The production of a significant quantity of propionic acid is the identifying feature of this genus when attempting to differentiate them from other gram-positive nonsporulating anaerobic bacteria. They can also produce some isovaleric acid, formic acid, succinic acid, and lactic acid.

All the species in this genus can use glucose to produce acid. They test positive for catalase. Species of *Propionibacterium* are distinguished using indole production, nitrate test, esculin hydrolysis, gelatin liquefication, and other biochemical tests such as the fermentation of sucrose, maltose, and mannitol.

The GC content of *Propionibacterium* DNA is 53–57% by Tm method. The type species is *P. freudenreichii*.

4.1.4.1 Propionibacterium acnes

P. acnes is a gram-positive irregular bacillus (Figure 4.4(A), (B), and (C)). It is either anaerobic or

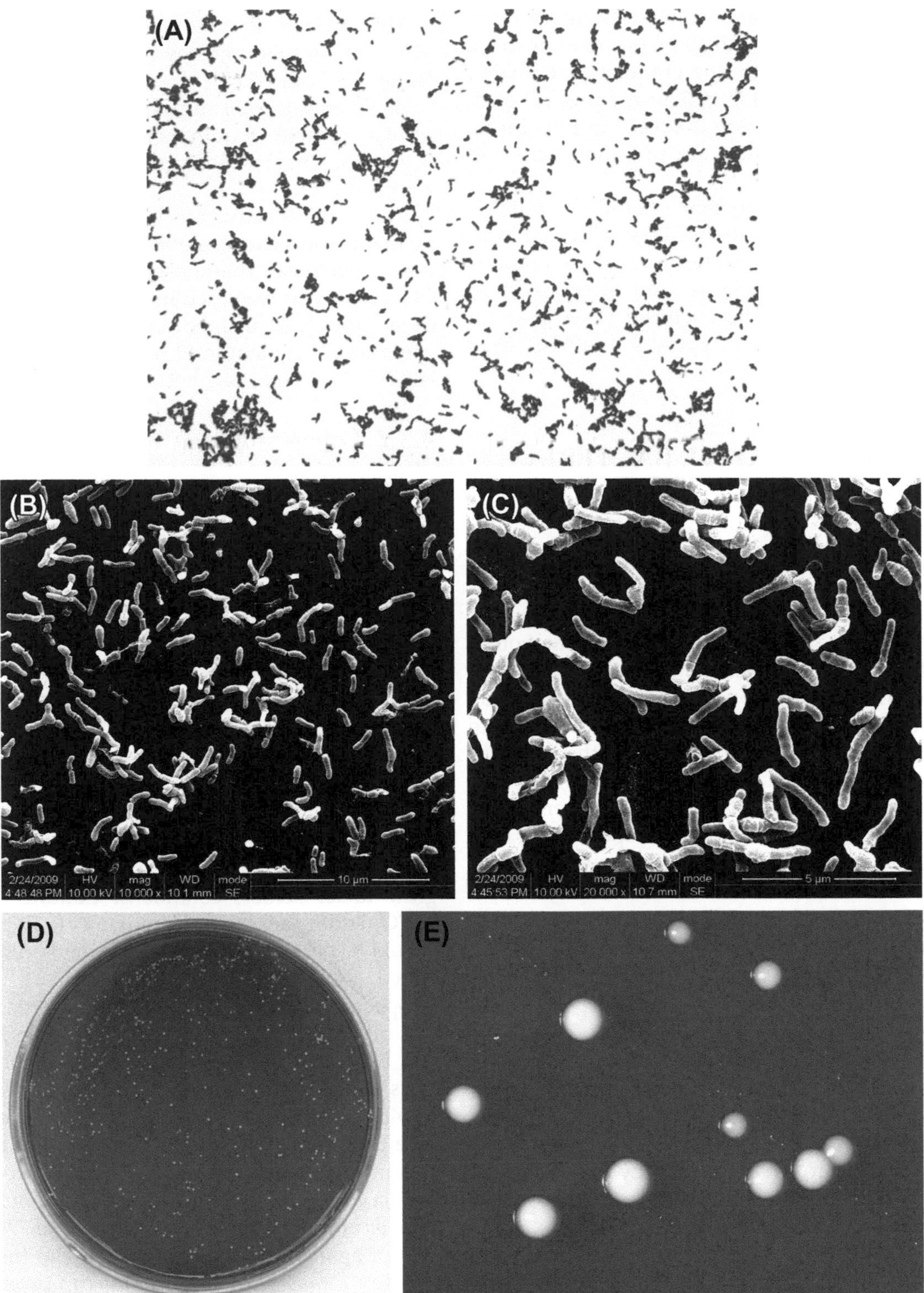

FIGURE 4.4 (A) *Propionibacterium acnes* cells (Gram stain). (B) *P. acnes* cells (SEM). (C) *P. acnes* cells (SEM). *P. acnes* cells are polymorphic but are mainly thin and long rod-shaped cells. Sero variant type II cells are mainly spherical with no spores. Cells stain gram-positive, but some can stain gram-negative. (D) *P. acnes* colonies (BHI blood agar). (E) *P. acnes* colonies (stereomicroscope). *P. acnes* can generate pinprick-sized colonies to colonies 0.5 mm in diameter, white or gray, glossy, translucent or opaque, pad-shaped or neat-edged colonies 2–3 days postinoculation on the surface of horse blood or rabbit blood agar. On rabbit blood agar, 68% of Sero variant type I strains can generate hemolytic reaction while Sero variant type II strains cannot.

microaerophilic. A medium with low redox potential is required for primary cultures. Characteristic colonies are shown in Figure 4.4(D) and (E). Culture in dextrose broth is cloudy or clear with fine granular precipitates. In old broth, the culture exhibits light red precipitation. The final pH value of a culture is 4.5–5.0 in PYG broth. The terminal acid metabolites are propionic acid and acetic acid.

P. acnes is part of the normal microflora, but the number of cells varies greatly between different individuals. *P. acnes* can be separated from the secretions from acne, wounds, blood, pus, and intestinal contents. The GC content of the bacterial DNA is about 57–60% (by Tm method). The type strain is ATCC6919 (NCTC737).

4.2 GRAM-NEGATIVE BACTERIA

4.2.1 *Bacteroides*

Bacteroides is a genus of gram-negative, obligate anaerobic, nonsporulating bacilli that belong to the family *Bacteroidaceae*. Members of this genus are chemoheterotrophs and can use carbohydrates, peptone, and other intermediate bacterial metabolites.

Bacteroides are naturally found in the mouth, tongue, intestinal tract, and vagina. All strains that can tolerate bile (20% oxgall salts) fall collectively into the *B. fragilis* group, including *B. fragilis*, *B. thetaiotaomicron*, etc. that are detected in cultures from appendicitis, peritonitis, and cervicitis clinical specimens. Other species that cannot tolerate bile and that produce melanin were reclassified to the genera *Porphyromonas* and *Prevotella*.

Cells measure 0.8–1.8 μm × 0.8–1.6 μm when grown in glucose broth, often showing visible vacuoles or darker stain at the ends. Cells are arranged singly or in pairs with no spores. Many strains are encapsulated.

4.2.1.1 Bacteroides fragilis

B. fragilis are gram-negative bacilli (Figure 4.5(A), (B), and (C)). They are obligate anaerobes and grow well in nutrient agar supplemented with chlorinated hemoglobin and vitamin K_1. Growth is inhibited by 20% bile. Optimum growth takes place at pH 7.0. Characteristic colonies are shown in Figure 4.5(D) and (E).

The cells mainly produce succinic acid and acetic acid in PYG liquid medium, but lactic acid, propionic acid, isovaleric acid, isobutyric acid, formic acid, benzene, and acetic acid are also produced. This species is biochemically active and is capable of fermenting glucose, lactose, maltose, fructose, raffinose, and other carbohydrates, hydrolyzing esculin. *B. fragilis* tests negative with the indole test.

The percent GC in *B. fragilis* DNA is 41–44% (using the Tm method). The type strain is ATCC25285 (NCTC9343).

4.2.1.2 Bacteroides thetaiotaomicron

This species is gram-negative (Figure 4.6(A) and (B)) and its morphology under SEM is shown in Figure 4.6(C) and (D).

B. thetaiotaomicron are obligate anaerobes and have similar growth requirements and characteristics as *B. fragilis*. Characteristic colonies are shown in Figure 4.6(E) and (F).

Cells exhibit active biochemistry and are able to ferment glucose, lactose, maltose, fructose, raffinose, arabinose, cellobiose, rhamnose, and other carbohydrates. Cells can hydrolyze esculin and test positive for indole. Cells grow well in medium containing 20% bile. The genomic GC content is 40–43% (by Tm). The type strain is ATCC29148 (NCTC10582).

4.2.2 *Capnocytophaga*

Capnocytophaga are gram-negative, facultative anaerobic bacteria. They were the earliest bacteria to be isolated and named from the human subgingival plaque. Common *Capnocytophaga* species are: *Capnocytophaga ochracer*, *Capnocytophaga sputigena*, *Capnocytophaga gingivalis*, *Capnocytophaga granulose*, and *Capnocytophaga heamolytica*.

Capnocytophaga cells are 0.42–0.6 μm × 2.5–2.7 μm in size and shaped like bent rods or filaments, usually with rounded or slightly pointed ends. The length of the cells varies. In liquid culture, cells are polymorphic or take on a long, filamentous morphology, and tight clumps can be observed. The bacteria produce no capsule and no sheath. They do not form spores, have no flagella, but have sliding motility.

Capnocytophaga are facultative anaerobes but do not grow under aerobic conditions. Cultures grow well in a CO_2-added anaerobic environment. Primary cultures should be performed in an aerobic environment with CO_2 added.

Species in this genus often form colonies of wet, thin, flat, diffuse growth with ragged edges on TS blood agar and BHI blood agar. After 24 h incubation at 35–37 °C, the size of colonies are like pinpricks. After incubation for 48–96 h, colonies become 2–4 mm in diameter and take on the appearance of bumps. Some colonies may become recessed into the agar. Aside from hemolytic *Capnocytophaga* (which produces β-hemolysis), other species are not hemolytic on blood agar. The concentration of agar in the medium affects the force of sliding motility. *Capnocytophaga* cultures can produce a special smell, similar to caramel or a bitter almond flavor.

Colonies on the agar surface can produce white to pink or orange-yellow pigmentation. Centrifuged cells appear to be an orange-yellow clump.

Capnocytophaga do not produce indole, can ferment glucose, lactose, maltose, mannose, and sucrose acid, and do not ferment mannitol and xylose. They can hydrolyze

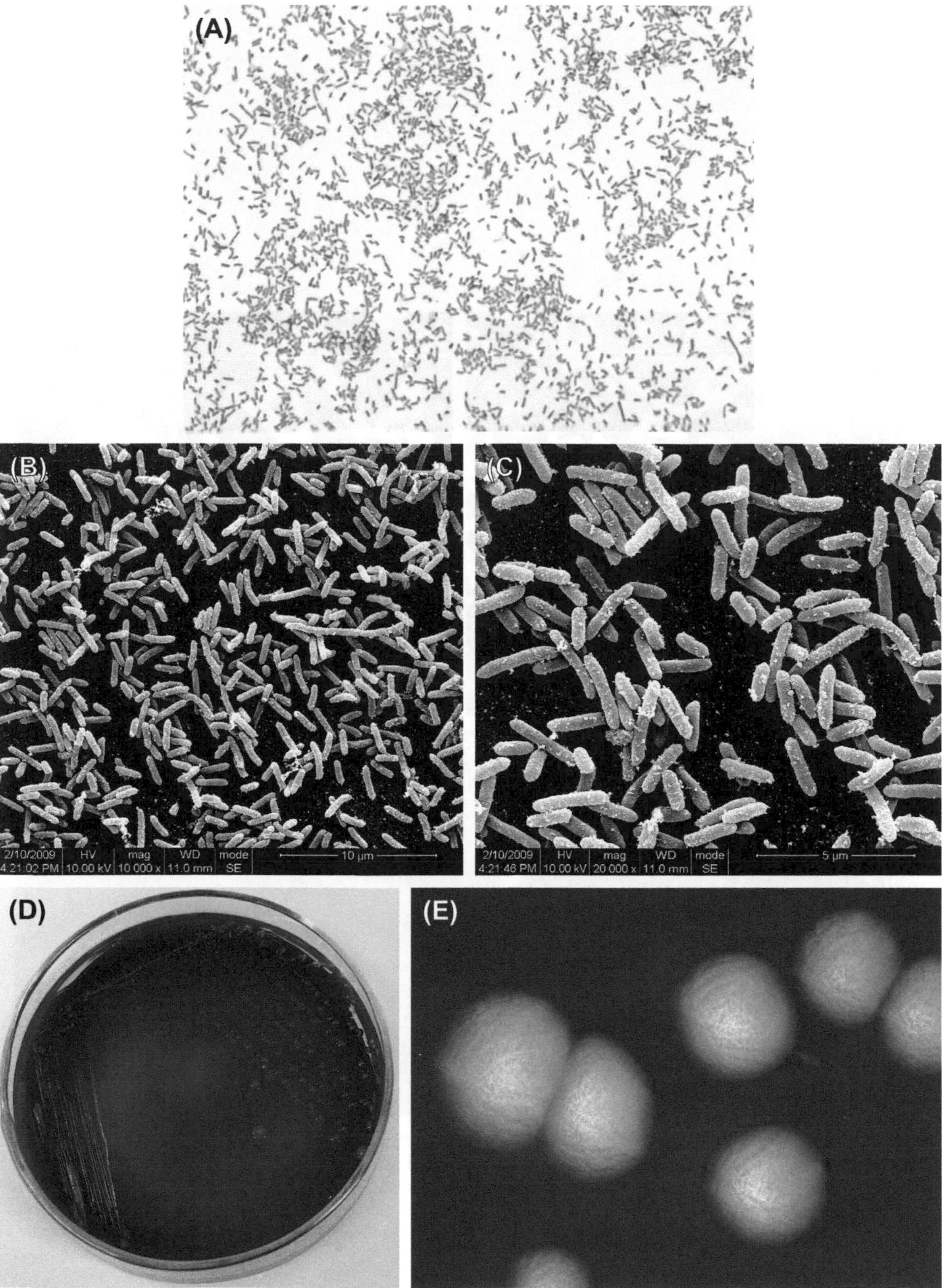

FIGURE 4.5 (A) *Bacteroides fragilis* (Gram stain). (B) *B. fragilis* (SEM). (C) *B. fragilis* (SEM). *B. fragilis* appears rounded on both ends, with visible vacuoles or darker stain at the ends. The cells are approximately 0.8–1.8 μm × 1.6–0.8 μm when grown in glucose broth. The cells can be single or in pairs, and they form no spores and are gram-negative. (D) *B. fragilis* colonies (BHI blood agar). (E) *B. fragilis* colonies (stereomicroscope). *B. fragili* can form gray colonies about 1–3 mm in diameter. The colonies are rounded, smooth, translucent or semitransparent on blood agar.

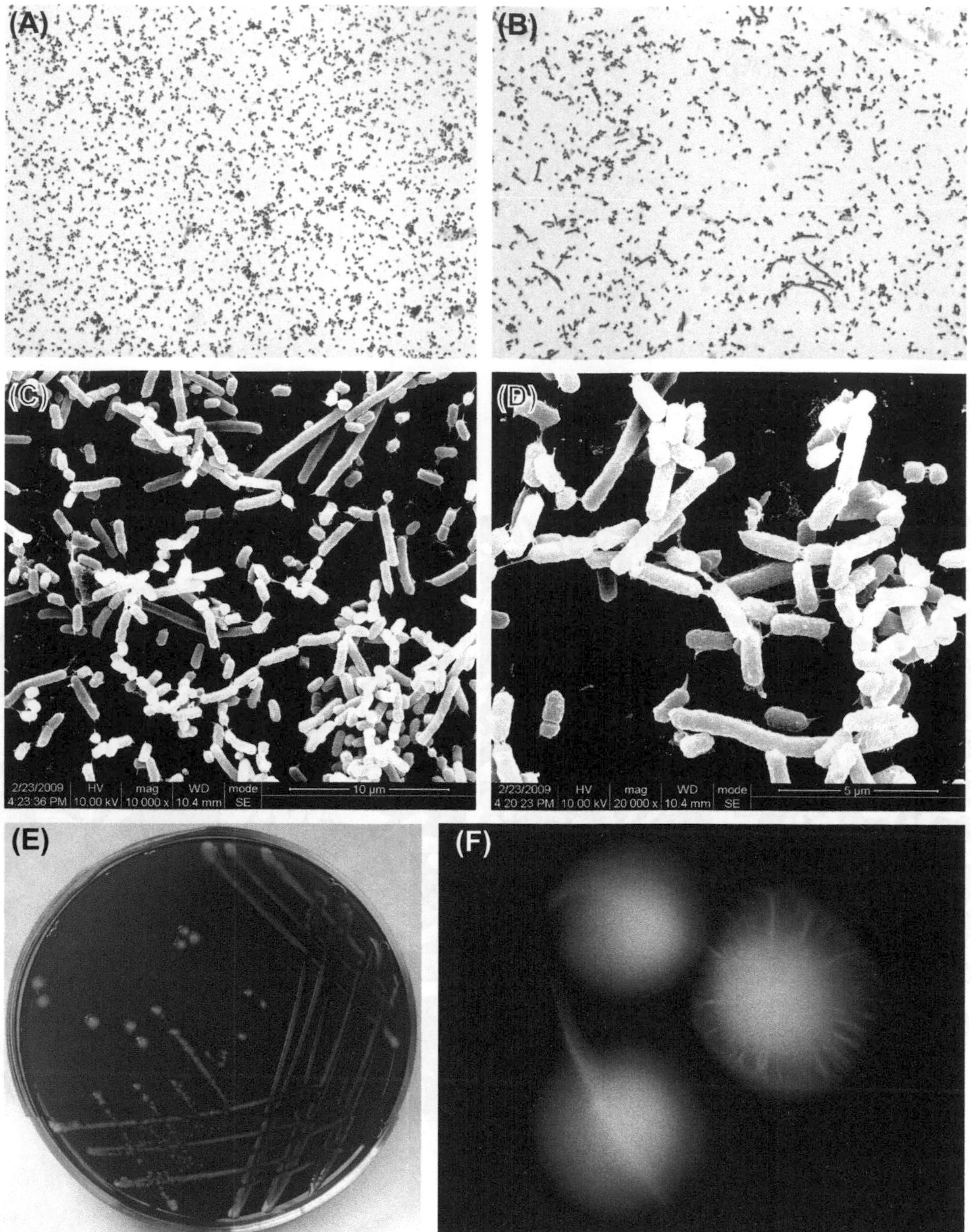

FIGURE 4.6 (A) *Bacteroides thetaiotaomicron* cells (Gram stain). (B) *B. thetaiotaomicron* cells (Gram stain). (C) *B. thetaiotaomicron* cells (SEM). (D) *B. thetaiotaomicron* cells (SEM). *B. thetaiotaomicron* appears spherical or rounded on both ends, with visible dark stain at the ends. Cells are 0.7–1.1 µm × 1.3–8.0 µm when grown in glucose broth, and are arranged as single cells or in pairs. Cells stain gram-negative. (E) *B. thetaiotaomicron* colonies (BHI blood agar). (F) *B. thetaiotaomicron* colonies (stereomicroscope). *B. thetaiotaomicron* colonies are about 1–3 mm in diameter. They form bumpy, glossy, soft, white, round colonies on blood agar. An orange peel-like appearance can be observed on the colony surface under stereomicroscope.

esculin and test negative for catalase and oxidase, while testing positive for ONPG and benzidine. Nitrate reduction, dextran hydrolysis, starch or gelatin hydrolysis and other biochemical tests can help identify this genus of bacteria.

A member of the normal microflora of humans and primates, this genus is mainly found to colonize the oral cavity. They are common oral bacteria and can be obtained from various parts of the oral cavity, including plaque, gingival sulcus, saliva and sputum, and throat specimens. These bacteria are often detected in mixed bacterial infections, such as juvenile periodontitis, infected root canal, dry socket after tooth extraction, oral ulcers, and other clinical specimens, and can also be

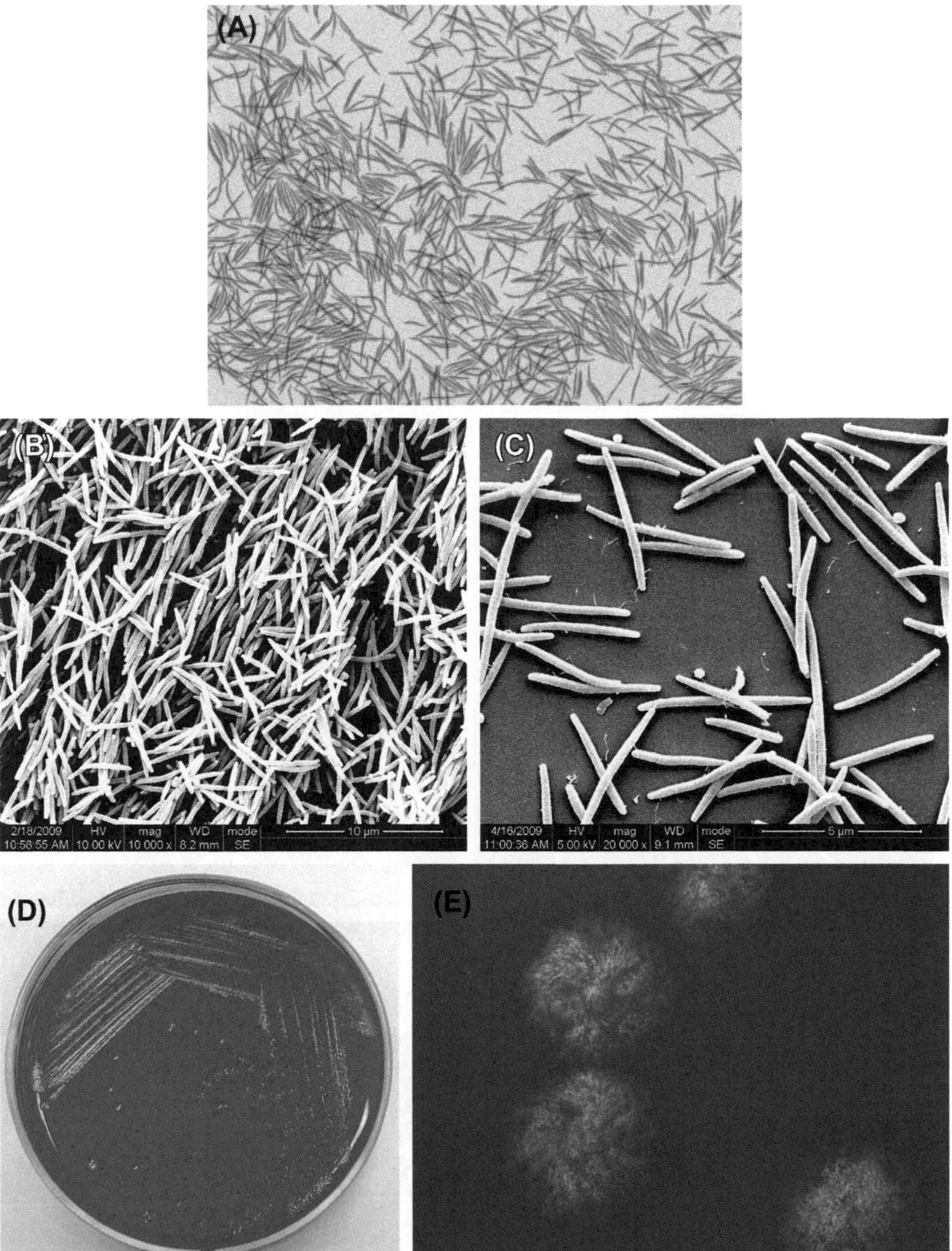

FIGURE 4.7 (A) Cells of *C. gingivalis* (Gram stain). (B) Cells of *C. gingivalis* (SEM). (C) Cells of *C. gingivalis* (SEM). Cells of *C. gingivalis* are fusobacterium-shaped, and the ends are usually rounded. Cells are often arranged in an orderly manner and stain negative by Gram stain. (D) Colonies of *C. gingivalis* (BHI blood agar). (E) Colonies of *C. gingivalis* (stereomicroscope). Colonies of *C. gingivalis* on BHI blood agar are irregular, gray colonies, with ragged edges. Typical hair-like diffuse colonies can be seen under the stereomicroscope.

isolated from bacteremia, soft tissue infections, injuries and abscesses at various locations, cerebrospinal fluid, vaginal, cervical, and amniotic fluid, trachea, and eyes.

The GC content of *Capnocytophaga* genomic DNA is 33–41% (by Tm method). The type species is yellowish *Capnocytophaga*.

4.2.2.1 Capnocytophaga gingivalis

This gram-negative bacterium is shown in Figure 4.7(A), (B), and (C). Characteristic colonies are shown in Figure 4.7(D) and (E).

C. gingvalis does not ferment lactose, galactose, amygdalin, salicin, cellobiose, esculin, and glycogen.

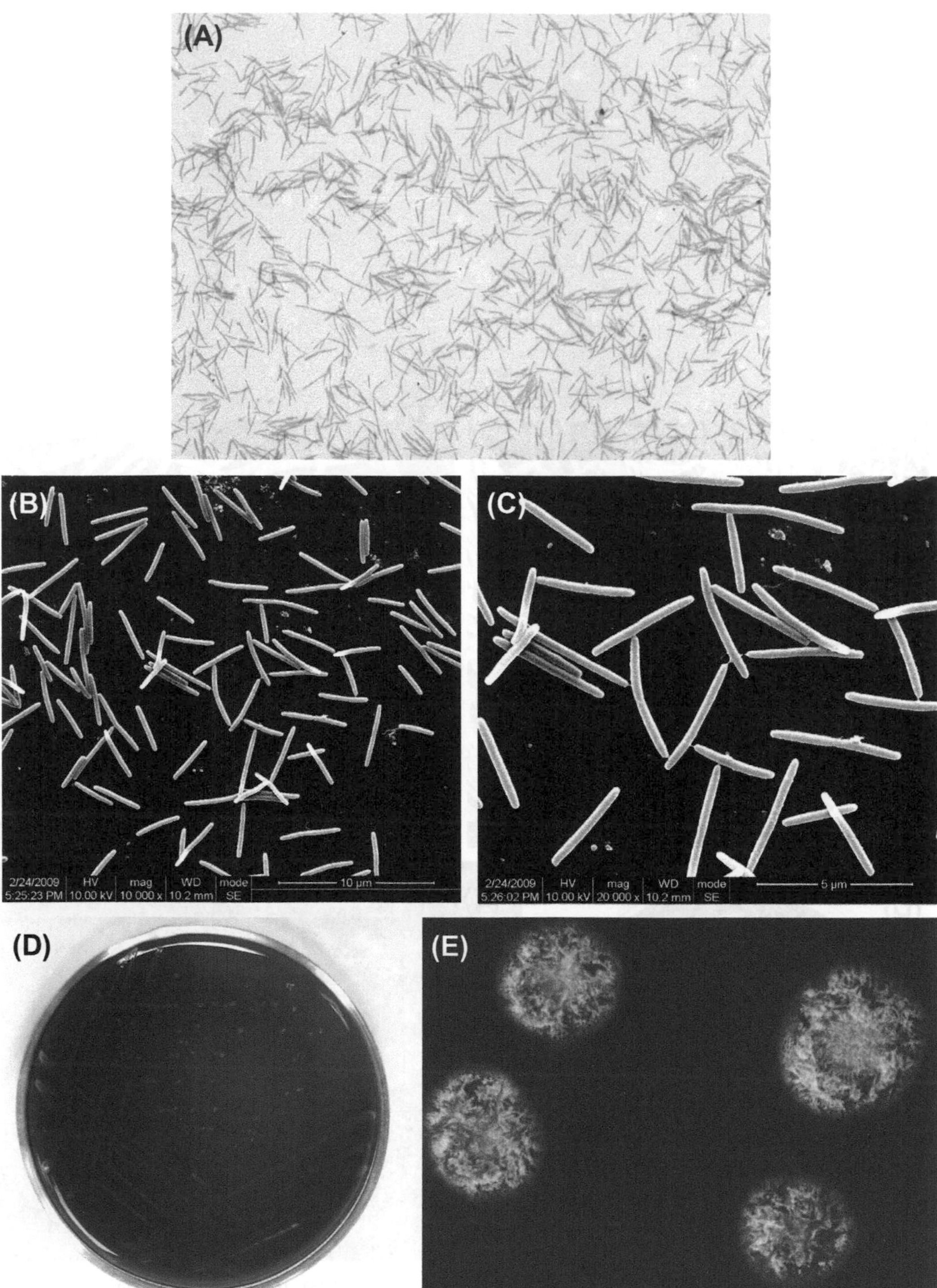

FIGURE 4.8 (A) Cells of *C. sputigena* (Gram stain). (B) Cells of *C. sputigena* (SEM). (C) Cells of *C. sputigena* (SEM). Cells of *C. sputigena* are bent bacilli, usually with rounded ends. They produce no spores and stain gram-negative. (D) Colonies of *C. sputigena* (BHI blood agar). (E) Mesh-like structure on the surface of colonies of *C. sputigena* (stereomicroscope). Colonies of *C. sputigena* on BHI blood agar surface are flat, spread orange colonies. Typical hair-like diffuse structure can be seen under the stereomicroscope.

It also does not hydrolyze starch, dextran, and gelatin. Only 8% of the strains can reduce nitrate.

The GC content in *C. gingivalis* DNA is 40% (by method). The type strain is ATCC33624.

4.2.2.2 Capnocytophaga sputigena

This species is a gram-negative bacillus (Figure 4.8(A), (B), and (C)). Characteristic colonies are shown in Figure 4.8(D) and (E).

The cells can ferment lactose, glucose, maltose, and sucrose but do not ferment mannitol, cellobiose, glycogen, and xylose. *C. sputigena* does not hydrolyze starch and dextran. The hydrolysis of gelatin and nitrate reduction are the most important features by which this species can be distinguished from other members of the genus. *C. sputigena* may be involved in juvenile periodontitis.

The GC content of *C. sputigena* DNA is 33–38% (Tm method). The type strain is ATCC33612.

4.2.3 *Eikenella*

Eikenella is a genus of gram-negative facultative anaerobic bacteria that do not produce spores.

4.2.3.1 Eikenella corrodens

E. corrodens was thus named because it produces typical colonies that can erode agar. It is also known as *Bacteroides corrodens* and is the only species in the genus *Eikenella*. Cells stain gram-negative (Figure 4.9(A), (B), and (C)).

E. corrodens is a facultative anaerobe, and primary cultures require anaerobic conditions or supplementation with 5–10% CO_2. It is essential to add hemin (5–25 mg/L) to the culture when grown under aerobic conditions. The optimum growth temperature is from 35 °C to 37 °C, the optimum pH is 7.3, and cultures require sufficient humidity. Characteristic colonies are shown in Figure 4.9(D), (E), and (F).

E. corrodens does not grow well in liquid media. Broth supplemented with 0.2% agar, cholesterol (10 mg/L), and 3% serum can promote its growth. Under aerobic conditions, 5–10% bile can inhibit growth. However, under anaerobic conditions, up to 10% bile can be tolerated.

E. corrodens is biochemically inactive. It does not ferment glucose and other carbohydrates or produce acid. It tests negative for catalase, urease, arginine dehydrogenase, and indole, but is positive for nitrate reduction, as well as oxidase and lysine decarboxylase.

E. corrodens is a member of the normal flora in the human oral cavity and intestinal tract. It can also be isolated from the upper respiratory tract and urogenital tract. As an opportunistic pathogen, it is often associated with other bacterial pathogens to cause mixed bacterial infections, especially in the mouth and respiratory tract. Its detection rate is higher in lesions of active adult periodontitis and specimens of dry socket after tooth extraction, and it is suspected to be related to periodontitis.

The GC content in its genomic DNA is 56–58% (by Tm method). The type strain is ATCC23834 (NCTC10596).

4.2.4 *Fusobacterium*

Fusobacterium is a group of gram-negative obligate anaerobic bacteria that do not form spores. They belong to the family *Bacteroidaceae*. Species mainly found in the oral cavity are *Fusobacterium nucleatum*, *Fusobacterium necrophorum*, and *Fusobacterium varium*.

Most *Fusobacteria* are spindle-shaped cells. They may also appear polymorphous. Polymorphic *Fusobacteria* can form globular or long filiform cells. *Fusobacterium necrophorum* can take on many other morphologies, including irregular spherical swollen cells and linear cells. The cells are nonmotile, do not form spores, and stain gram-negative.

These obligate anaerobes can be grown in aerobic conditions on agar plates when 5–10% CO_2 is added to the culture conditions. Their sensitivity to oxygen depends on the specific bacterial species, the quantity of cells inoculated, and the type of culture medium.

Fusobacterium can be detected in clinical specimens of pus or gangrene infections. *Fusobacterium nucleatum* has a high prevalence in saliva and dental plaque, and is considered to be one of the bacteria involved in mixed infections of periodontitis, root canal infection, and postextraction infection.

The genomic GC content of this genus is 26–34% (Tm). The type species is *F. nucleatum*.

4.2.4.1 Fusobacterium nucleatum

The species *F. nucleatum* contains five subspecies: *F. nucleatum* animal subspecies (*F. nucleatum* subsp. *animalis*), fusiform subspecies (*F. nucleatum* subsp. *fusiforme*), nuclear subspecies (*F. nucleatum* subsp. *nucleatum*), polymorphic subspecies (*F. nucleatum* subsp. *polymorphum*), and Vincent subspecies (*F. nucleatum* subsp. *vincentii*).

The cells are gram-negative spindle-shaped coli. Cell morphologies are shown in Figure 4.10(A), (B), and (C). *F. nucleatum* is an obligate anaerobe, but can grow under atmospheric conditions with >6% oxygen by volume. The cells are viable even after 100 min of exposure to the air. Characteristic colonies are shown in Figure 4.10(D), (E), and (F).

F. nucleatum is not biochemically active. They do not transform lactate into propionate. They can produce indoles and DNase, but do not produce phosphatase. Most of the strains produce H_2S and can agglutinate red blood cells from both humans and animals.

F. nucleatum is mainly found on transparent gingiva and the gingival groove, and make up the oral normal flora. They can also be isolated from upper respiratory tract and chest infections, and occasionally from wounds and other sites of infection. *F. nucleatum* has a high rate of detection in destructive periodontal disease and

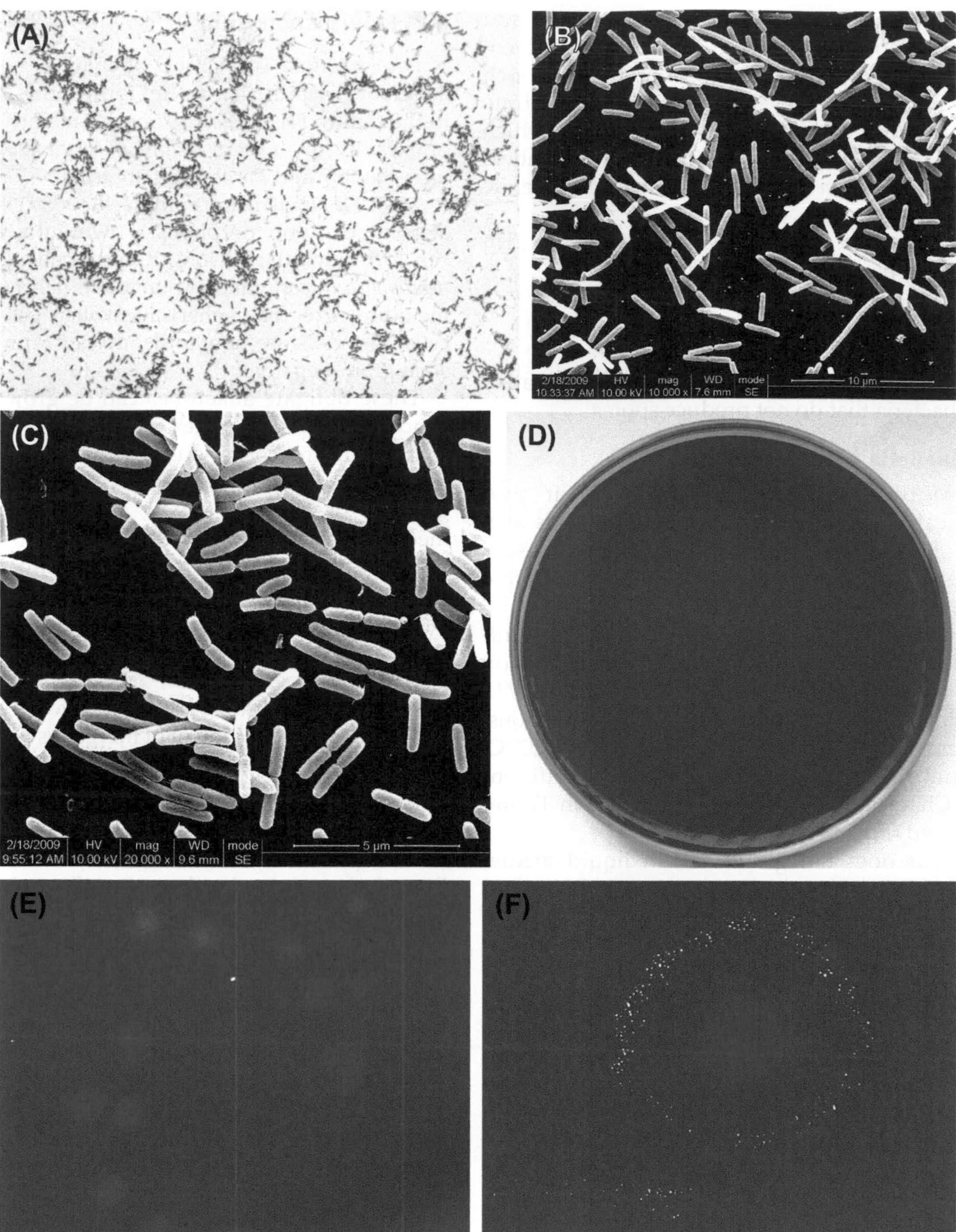

FIGURE 4.9 (A) Cells of *E. corrodens* (gram-negative). (B) Cells of *E. corrodens* (SEM). (C) Cells of *E. corrodens* (SEM). Cells of *E. corrodens* are 0.3–0.4 μm × 1.5–4.0 μm in size and rounded at the ends. They are mostly rod-shaped, short rod-shaped, or club-shaped, and can sometimes be found in the shape of a short wire rod. The cells do not produce spores and are nonmotile. "Tremor-shaped movement" can be seen on the surface of the agar. Cells stain gram-negative. (D) Colonies of *E. corrodens* (BHI blood agar). (E) Colonies of *E. corrodens* (stereomicroscope). (F) A pearlescent ring can be observed at the center of colonies of *E. corrodens* (stereomicroscope). Agar cultures of *E. corrodens* have a bleach-like smell, similar to *Haemophilus* and *Pasteurella* cultures on agar. Colonies do not produce hemolytic reactions on blood agar, but a light green ring can be seen around colonies. Two different types of colonies can form on blood agar: invasive phenotype and noninvasive phenotype. The invasive strain forms on the surface of blood agar when conditions are 36 °C, with 15% CO_2 and 100% humidity. Colony diameter ranges from 0.2 mm to 0.5 mm (after 24 h culture) or from 0.5 mm to 1.0 mm (after 48 h culture). Colonies are light yellow and opaque, and the center of the colony has a clear pearlescent ring. The edge of the colony is rough, refractive, and has a hair-like diffuse edge, and "tremor-shaped movement" can be seen on the surface of the agar. The noninvasive phenotype forms colonies with a diameter of 0.5–1 mm. The colonies are hemispherical, translucent, with no hair-like diffuse edge. They do not invade agar, show no adhesion to the agar, and have no "tremor-shaped movement."

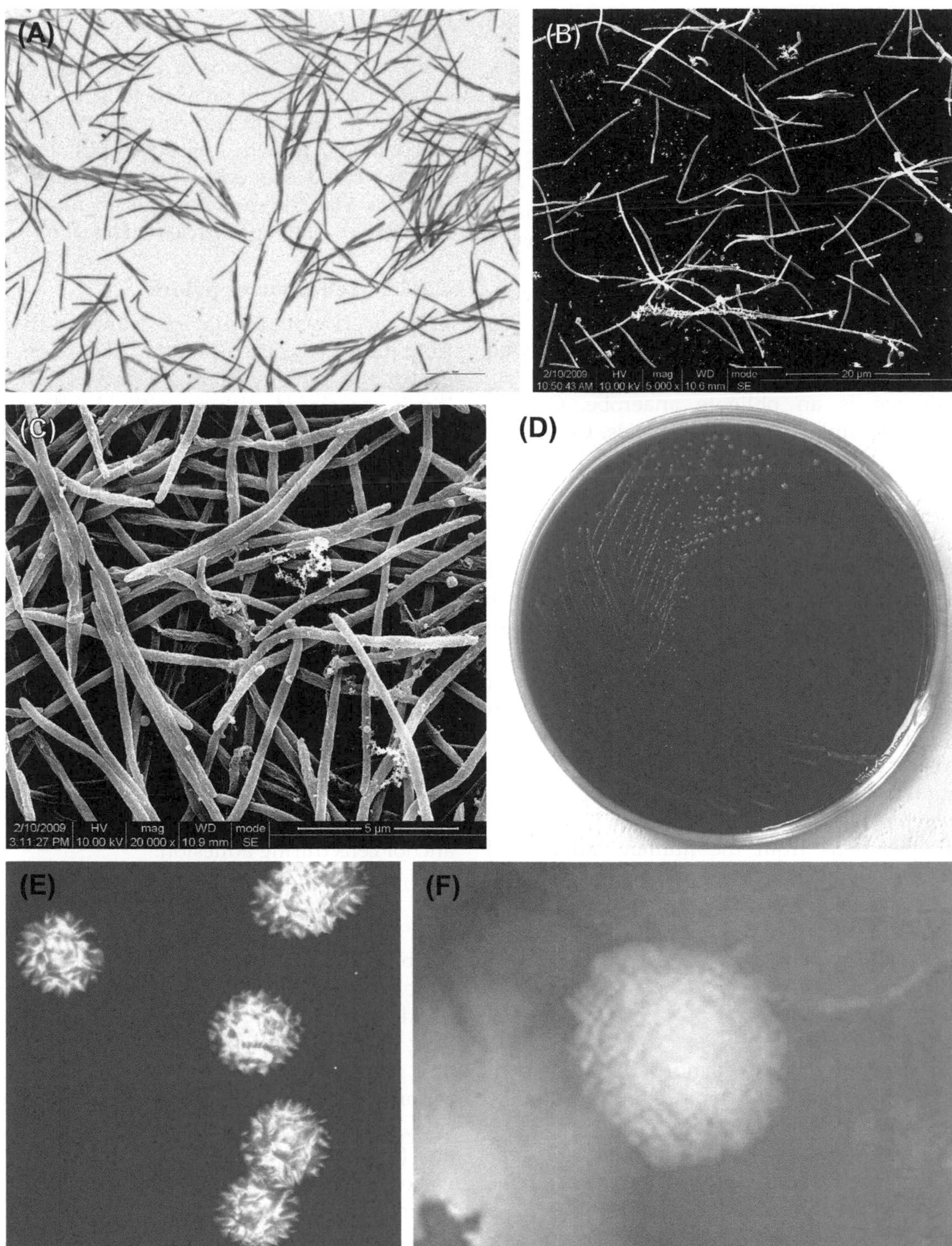

FIGURE 4.10 (A) *Fusobacterium nucleatum* cells (Gram stain). (B) *F. nucleatum* cells (SEM). (C) *F. nucleatum* cells (SEM). *F. nucleatum* measures 0.4–0.7 µm × 3–10 µm in diameter in glucose broth culture, with a fusiform or tapered end. Swelling can often be observed at the center of the cell, and gram-positive particles can be observed inside the cell. Cell length is generally associated with growth conditions. They have no pili and no flagella. Cells stain gram-negative. (D) *F. nucleatum* colonies (BHI blood agar). (E) *F. nucleatum* "breadcrumb" colonies (stereomicroscope). (F) *F. nucleatum* isolated from saliva samples (stereomicroscope). *F. nucleatum* can form colonies 1–2 mm in diameter. Colonies are rounded or slightly irregular, bumpy, pad-shaped, translucent, and with odor on blood agar plates. Colonies commonly appear to have spots of light that shine through. These are referred to as "breadcrumb" colonies. *F. nucleatum* generally do not produce hemolytic reactions on horse and rabbit blood agar. There is visible flocculent or particle precipitation in glucose broth cultures, but the broth does not necessarily become turbid. The culture has a foul smell and the final pH value of a glucose broth culture is pH 5.6–6.2.

infectious dental pulp, as it assists other pathogens in establishing oral infectious diseases.

The genomic GC content is 27–28%. The type strain is ATCC25586.

4.2.4.2 Fusobacterium necrophorum

There are two subspecies of *F. necrophorum*: *F. necrophorum* fundamental form subspecies (*F. necrophorum* subsp. *funduliforme*) and *F. necrophorum* necrosis subspecies (*F. necrophorum* subsp. *necrophorum*).

F. necrophorum is a gram-negative bacillus, with diverse cellular morphologies (Figure 4.11(A), (B), (C), (D), and (E)). Heptose and KDO are among the cell wall lipopolysaccharides.

F. necrophorum is an obligate anaerobe. Culture requirements are similar to related species. Characteristic colonies are shown in Figure 4.11(F), (G), (H), (I), and (J).

When cultured in glucose broth medium, *F. necrophorum* appears muddy, and smooth, flocculent particles or filamentous precipitation can also be found. The final pH value of a culture in glucose or fructose broth is 5.6–6.3. A few strains have a final pH value of 5.8–5.9 in maltose cultures.

F. necrophorum can agglutinate red blood cells from human, rabbit, and guinea pig blood, but not blood from cattle. It cannot hydrolyze glucan. Neither phosphatase, superoxide dismutase, nor lysine decarboxylase is detected, but it can produce DNase.

F. necrophorum is mainly isolated from some clinical disease specimens from the human and animal body, including abscesses, blood and necrotic lesions, and especially liver abscesses. The bacteria can also be detected in the mouth.

The genomic GC content is 31–34%. The type strain is ATCC25286.

4.2.4.3 Fusobacterium varium

This gram-negative species' cell morphology is shown in Figure 4.12(A), (B), and (C). Heptose and KDO make up its cell wall lipopolysaccharides.

F. varium is an obligate anaerobe. Its culture conditions are similar to other related species. Characteristic colonies are shown in Figure 4.12(D) and (E).

F. varium cannot hydrolyze glucan and does not produce phosphatase, but it does produce lysine dehydrogenase and DNase.

F. varium can be isolated from the human mouth and human feces, it can also be detected in suppurative infectious wounds, the upper respiratory tract, and in peritonitis, but its pathogenicity is not yet well defined.

The genomic GC content is 26–28%. The type strain is ATCC8501 (NCTC10560).

4.2.5 *Helicobacter*

Helicobacter is a genus of gram-negative, curved rod-shaped bacteria with polar flagellae. Because of the differences in their ultrastructure, fatty acid composition, morphology, growth characteristics, enzyme activity, and 16s rRNA sequence when compared to the genus *Campylobacter*, the species in this genus were classified as *Helicobacter*. The type species is *H. pylori*.

4.2.5.1 Helicobacter pylori

H. pylori was first isolated and cultivated in vitro from the gastric mucosal tissue of patients with chronic gastritis in 1983 by Marshall and Warren. The organism is associated with gastritis, duodenal and gastric ulcers, and gastric cancer, as well as a number of diseases at distant sites. The World Health Organization/International Agency for Research on Cancer (WHO/IARC) listed *H. pylori* as a Class A carcinogen in 1994.

H. pylori are nonsporulating, slender, curved rods measuring approximately 2.5–4.0 μm × 0.5–1.0 μm. They are pleiomorphic and can typically be found as spiral, s-shaped or gull-shaped cells (Figure 4.13(A), (B), and (C)). Cells stain gram-negative with one or more flagella. Sometimes rods or coccoid forms can be found in addition to the typical cellular morphologies when grown on solid culture medium. Under electron microscope, the bacteria have two to six flagella that are approximately 30 nm in thickness and 1–1.5 times longer than the bacterial cell. Flagella play a role in cellular movement and anchor cells during adhesion.

H. pylori grow under microaerophilic conditions and have high nutritional requirements. They grow well on Columbia blood agar medium containing 5% defibrinated blood or on brain heart infusion blood agar at 37 °C, 95% humidity, and under 10% carbon dioxide, 5% oxygen, and 85% nitrogen. Because of the long growth period, primary cultures require 3–7 days, and subcultures need 2–4 days to grow. Colonies are pinprick-sized, circular, neat, raised, colorless, translucent, and measure approximately 0.5–1 mm in diameter (Figure 4.13(D) and (E)). Resistance of *H. pylori* to various stressors is weak, as they survive for less than 3 h in air and no more than 1 day at 4 °C. Cultures are sensitive to heat, and the only way to preserve cells for the long term is by cryopreservation at −80 °C.

H. pylori is not biochemically active. It usually does not produce acid from sugar. Although there have been a few reports of acid production, it only becomes apparent in media containing a low concentration of peptone. Sugar is not usually the sole carbon source.

H. pylori cellular vitality significantly weakens at pH 3.5 and below, while physiological concentrations of bile salts can inhibit culture growth. *H. pylori*

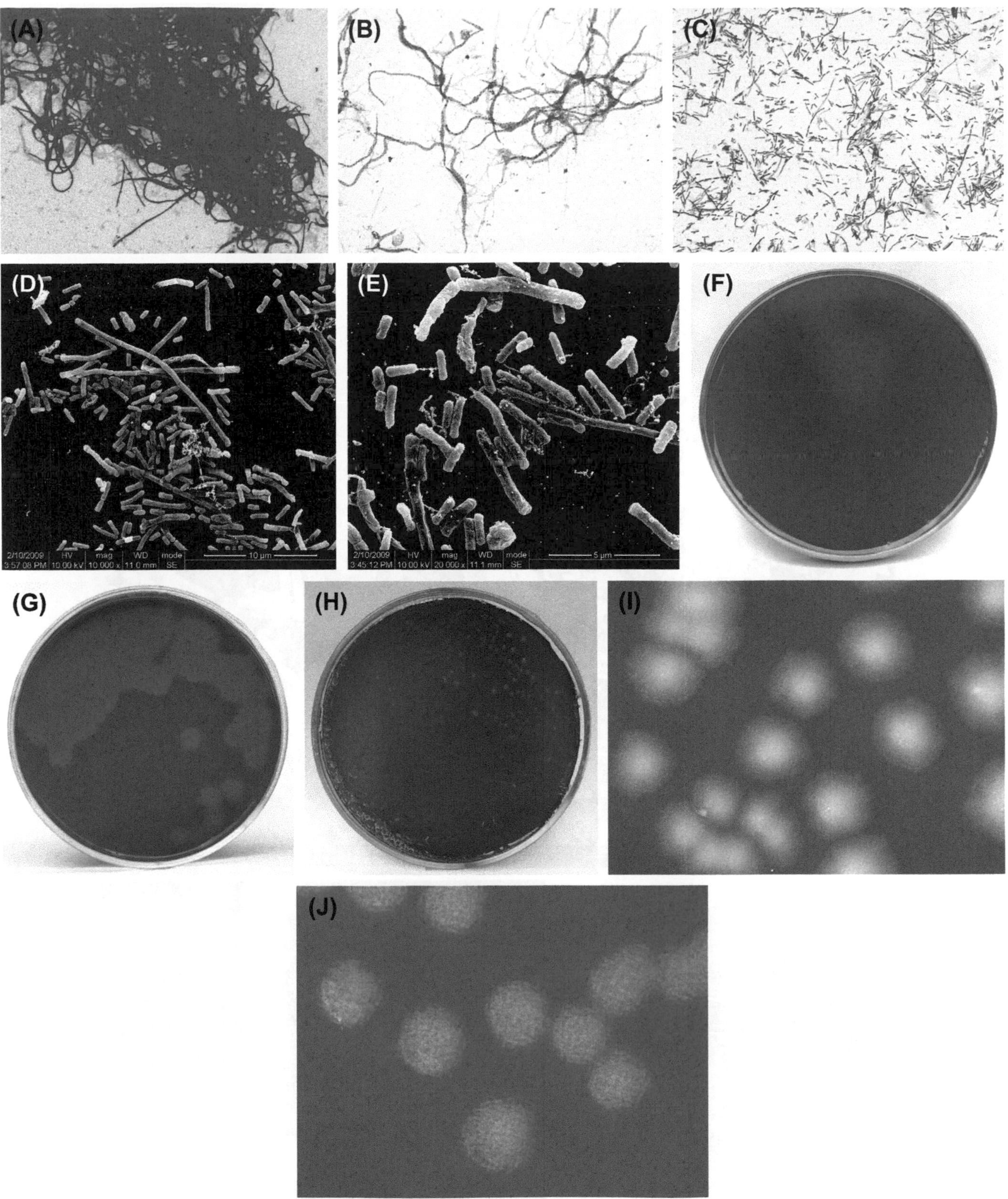

FIGURE 4.11 (A) Filament of *F. necrophorum* (Gram stain). (B) Filament of *F. necrophorum* with particles content (Gram stain). (C) *F. necrophorum* bacillus (Gram stain). (D) *F. necrophorum* cells (SEM). (E) *F. necrophorum* cells (SEM). *F. necrophorum* is morphologically diverse. The cell length can vary significantly and can be spheroids, rod-shaped to filaments more than 100 μm long. Although cells are approximately 0.5–0.7 μm in diameter, when cultured in glucose broth, they may swell to more than 1.8 μm in length, with round or pointed ends. Filamentous material containing particles are commonly seen in broth cultures, while coliform cells are found in old cultured cells or on agar. Cells stain gram-negative. (F) *F. necrophorum* colonies producing α-hemolysis (BHI blood agar). (G) *F. necrophorum* colonies producing β-hemolysis (BHI blood agar). (H) *F. necrophorum* colonies (BHI sheep-blood agar). (I) *F. necrophorum* zigzag colonies with mosaic internal structure (BHI sheep-blood agar, stereomicroscope). (J) *F. necrophorum* colonies with thread structure (stereomicroscope). *F. necrophorum* forms colonies 1–2 mm in diameter, rounded, raised above the surface, cream colored, translucent to opaque, with a fan-like or serrated edge on blood agar. A mosaic internal structure can be seen in the light. Most strains can produce α- or β-hemolysis on rabbit blood agar. In general, β-hemolytic strains are positive for lipase (on egg yolk agar), while α-hemolytic and nonhemolytic strains are negative for lipase and do not produce lecithin enzyme.

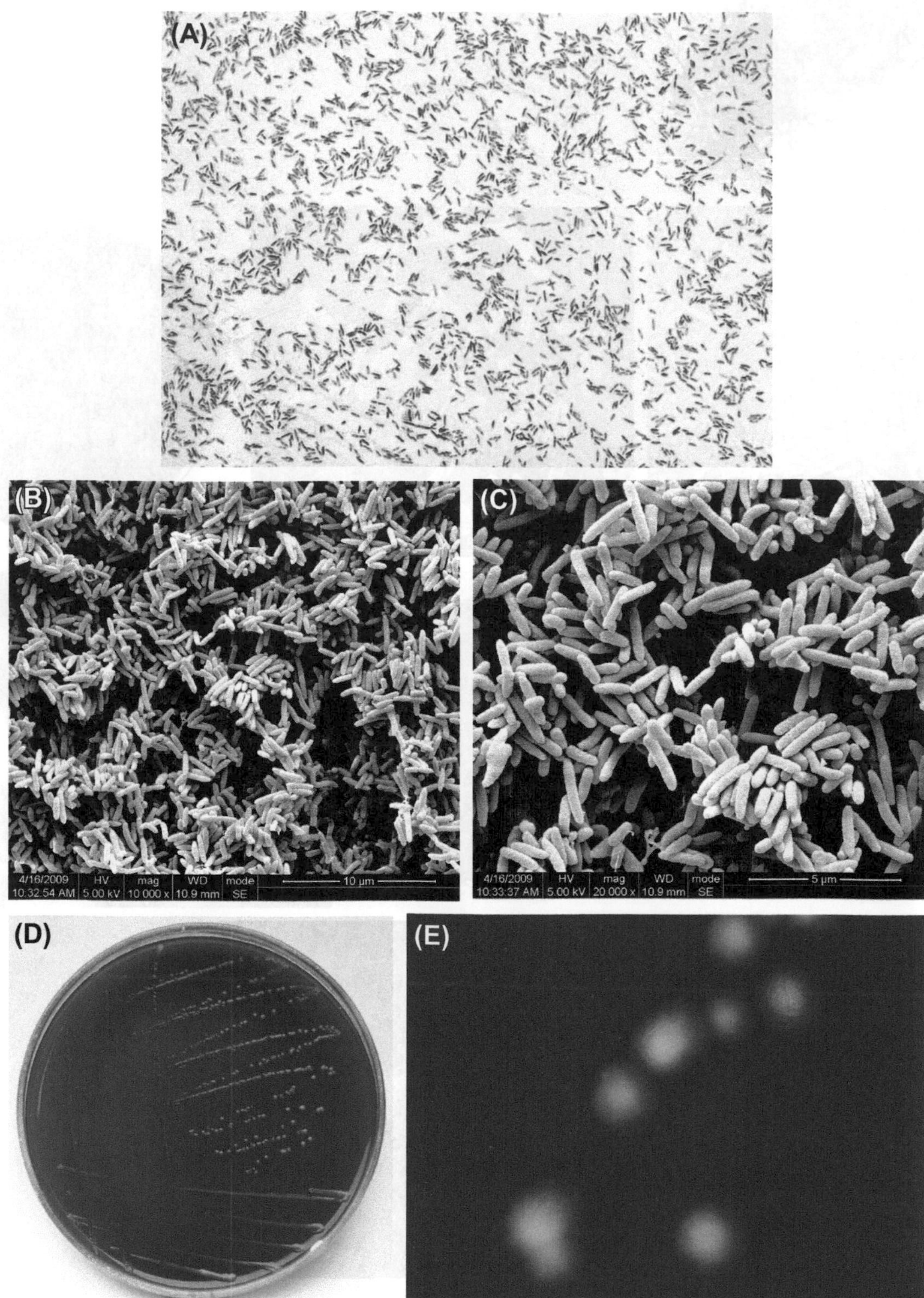

FIGURE 4.12 (A) *Fusobacterium varium* cells (Gram stain). (B) *F. varium* cells (SEM). (C) *F. varium* cells (SEM). *F. varium* cells measure 0.3–0.7 μm × 0.7–2.0 μm. They are polymorphic. Both cocci and bacilli can be observed and cells are present as single cells or in pairs. Cells stain gram-negative but also show uneven staining. (D) *F. varium* colonies (BHI blood agar). (E) *F. varium* colonies (stereomicroscope). *F. varium* forms pinprick-sized colonies to colonies 1 mm in diameter. They appear rounded, low, flat, convex, with a gray center and a colorless, translucent, and neat edge on blood agar.

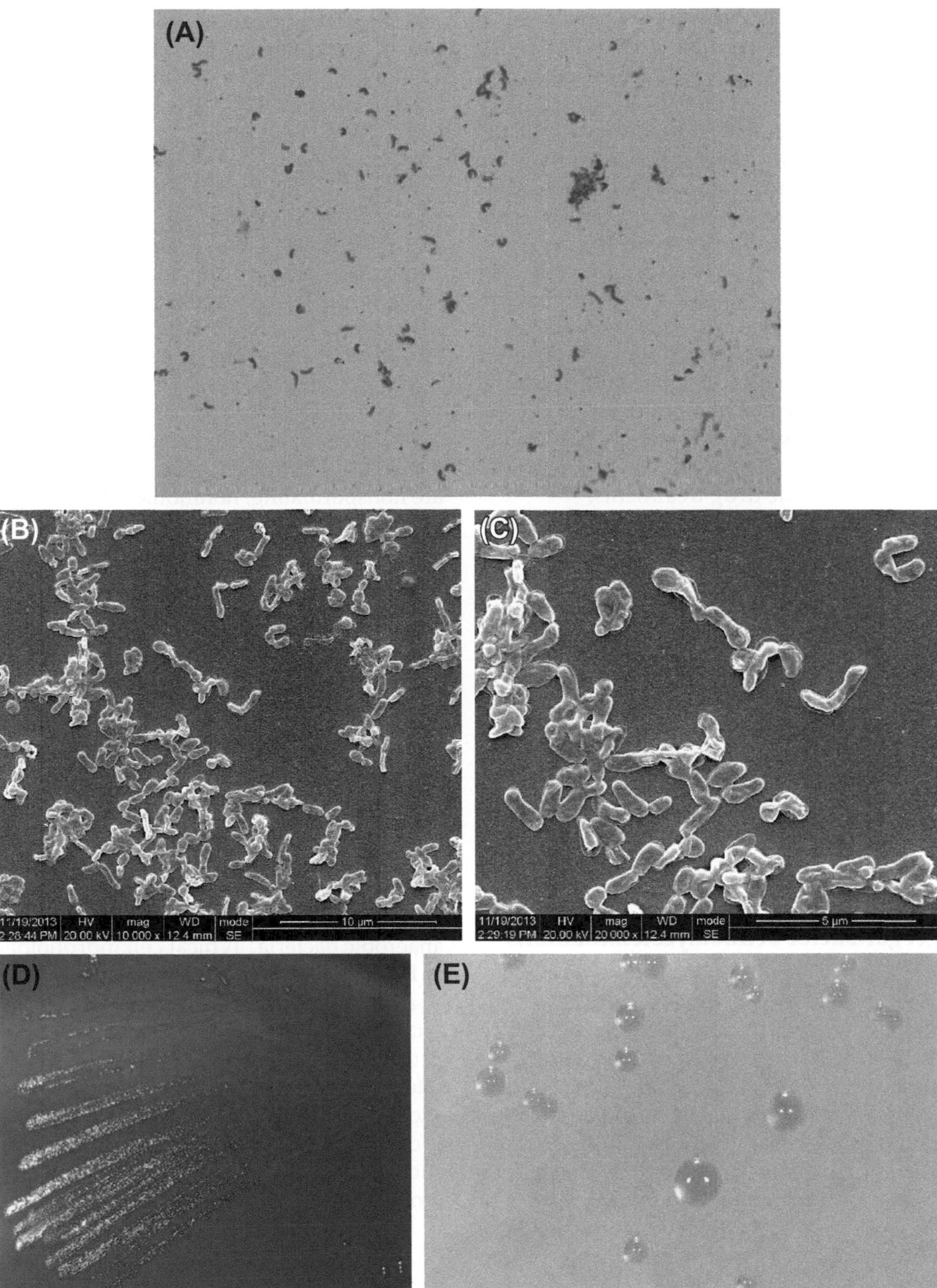

FIGURE 4.13 (A) *Helicobacter pylori* cells (Gram stain). Gram-negative cells are shaped like curved rods, S-shaped, or gull-shaped. (B) *H. pylori* cells (SEM). (C) *H. pylori* cells (SEM). (D) *H. pylori* colonies (BHI blood agar). Tiny pinprick colonies, circular, neat, and raised, colorless, translucent, about 0.51 mm in diameter and almost do not produce hemolytic reaction. (E) *H. pylori* colonies (stereomicroscope).

produces a large quantity of highly active extracellular urease, producing more than 400 times the urease activity than *Proteus*. *H. pylori* can also produce oxidase, catalase, alkaline phosphatase, γ-GGTP, etc. Seven positive enzymatic reactions are used as the basis for *H. pylori* biochemical identification. These are oxidase, catalase, urease, alkaline phosphatase, γ-GGTP, and leucine peptidase.

The infection rate with *H. pylori* is over 50% worldwide. The infection does not resolve itself and is sensitive to antibiotic treatment, but the recurrence rate is high. The oral cavity is considered to be a secondary site of colonization of *H. pylori* and may be associated with its high recurrence rate. *H. pylori* has been found in supragingival plaque or subgingival plaque and saliva, decayed teeth, infected root canals, and mucous membranes of the tongue, buccal and palatal mucosa by a large number of researchers using various molecular technologies. Isolating *H. pylori* using the traditional culturing method is particularly difficult, as there are many different kinds of bacteria in the oral cavity that form the dental plaque biofilm. The complex mutually beneficial relationship of coexistence and competition that is present within the biofilm makes for a relatively stable environment resistant to external stimuli. As a result, oral *H. pylori* can more easily evade drug treatments.

Oral *H. pylori* infections may be related to oral infectious diseases such as periodontal disease, tooth decay, oral ulcer, etc. Studies confirmed that periodontal disease is more likely to arise in *H. pylori*-positive patients, and the rate of infection in *H. pylori*-positive patients is positively correlated with the degree of inflammation. Triple therapy combining basic periodontal treatment (supragingival and subgingival scaling) can enhance the eradication of *H. pylori* and reduce the recurrence rate of *H. pylori*.

The genome size of *H. pylori* is about 1.67×10^6 bp and the GC content is 37%. The type strain is ATCC43504.

4.2.6 *Aggregatibacter*

4.2.6.1 Aggregatibacter actinomycetemcomitans

Formerly classified as *Actinobacillus actinomycetemcomitans*, this species was subsequently named *Haemophilus actinomycetemcomitans*, and is now known as *Aggregatibacter actinomycetemcomitans*. It is a major pathogen in juvenile periodontitis.

These are gram-negative bacteria, and the morphology of bacterial cells is shown in Figure 4.14(A), (B), and (C). *A. actinomycetemcomitans* is a facultative anaerobe and grows well in the microaerophilic environment of 5–10% CO_2. Its optimum growth temperature is 37°C and it does not grow at 22°C. Characteristic colonies are shown in Figure 4.14(D), (E), and (F).

Cells ferment fructose, glucose, maltose, and mannose and produce acid, but do not ferment sucrose, trehalose, lactose, raffinose, melibiose, and arabinose. *A. actinomycetemcomitans* tests positive for oxidase and catalase, can reduce nitrate, does not hydrolyze esculin and hippurate sodium, and does not produce H_2S and indole.

The main colonization site of this species is subgingival plaque. It is detected both in normal oral bacteria and also in lesions of juvenile periodontitis patients at a higher detection rate. Therefore, *A. actinomycetemcomitans* is considered an important pathogen.

The genomic GC content is 43% (by Tm method). The type strain is NCTC9710.

4.2.7 *Prevotella*

Prevotella is a genus named after the French microbiologist A. R. Prevol. These bacteria belong to the genus *Bacteroides* and include bile-sensitive strains and melanin-producing, sugar metabolizing strains.

The main species of *Prevotella* found in the oral cavity are: *Prevotella intermedia*, *Prevotella melaninogenica*, *Prevotella loescheii*, *Prevotella nigrescens*, *Prevotella dentocola*, and *Prevotella corporis*.

The most frequently found cell morphology is short bacillus, but long bacilli are sometimes observed. The cells are nonsporulating, nonmotile, and stain gram-negative.

Bacteria belonging to this genus are obligate anaerobes, and most cultures require supplementing with hemin and vitamin K. On blood agar, *Prevotella* spp. produce melanin to form black colonies. In PYG liquid medium, these bacteria can produce acetic acid and succinic acid as the major terminal acid products. Species of this genus ferment glucose and hydrolyze gelatin. They are sensitive to bile salt and can thus be distinguished from other bacteroides that can tolerate bile salts.

Prevotella are dominant bacteria in the human gingival groove and are also the suspected pathogens behind periodontitis.

4.2.7.1 Prevotella intermedia

P. intermedia is a gram-negative, black-pigmented, anaerobic bacteria. Typical cell morphologies are shown in Figure 4.15(A), (B), and (C).

As an obligate anaerobe, cultures require hemin and vitamin K. Most species grow well in temperatures between 25°C and 45°C. For other cultural demands, refer to the genus *Prevotella*. Characteristic colonies are shown in Figure 4.15(D) and (E). Culture in glucose broth is cloudy, with even precipitation and sticky or slightly sticky deposits at times. The final pH value of glucose broth ranges from pH 4.9 to 5.4.

P. intermedia can produce indole and hydrolyze gelatin, but not esculin. Cells ferment glucose and sucrose, but not arabinose, larch sugar, cellobiose, rhamnose, galactose, and salicin. Cells can produce superoxide dismutase, but do not hydrolyze dextran.

The site of colonization is the gingival sulcus, but cells also can be found in saliva, dental calculus, and other plaque specimens. Previous research has shown that this species is the main pathogenic bacteria in pregnancy-related gingivitis. In addition, the bacteria can also be isolated from pericoronitis, the focus of infection after tooth extraction, infected root canals, infections of the

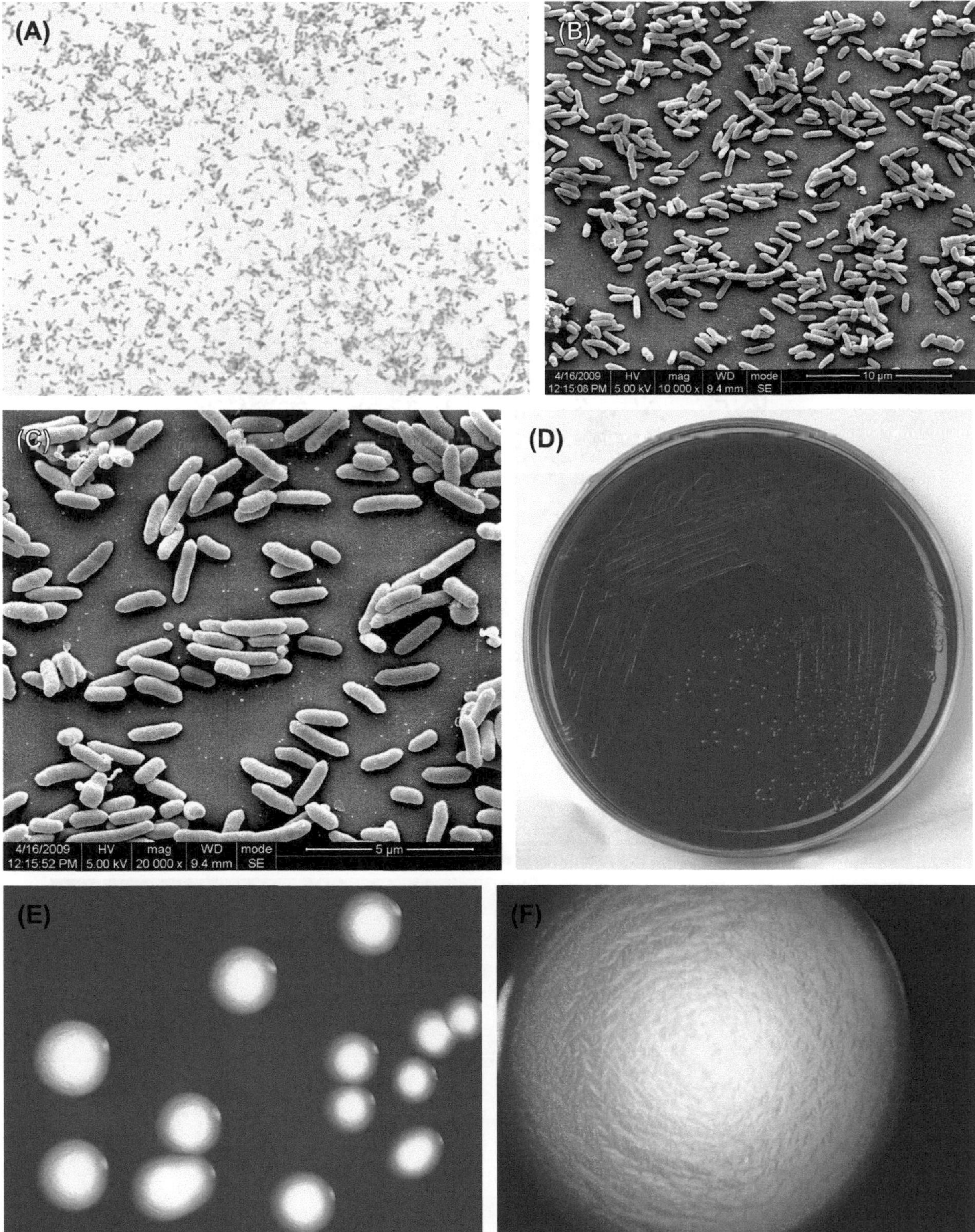

FIGURE 4.14 (A) Cells of *A. actinomycetemcomitans* (Gram stain). (B) Cells of *A. actinomycetemcomitans* (SEM). (C) Cells of *A. actinomycetemcomitans* (SEM). *A. actinomycetemcomitans* are 0.5–0.8 µm × 0.6–1.4 µm in size. Cells are spherical, club-shaped, or rod-shaped. Rod-shaped cells are common in agar cultures. The cells arrange as single cells, in pairs, or in piles. They produce no spores, are nonmotile, and do not form capsules. Cells stain gram-negative. (D) *A. actinomycetemcomitans* colonies (BHI blood agar). (E) *A. actinomycetemcomitans* colonies (stereomicroscope). (F) *A. actinomycetemcomitans* colonies (stereomicroscope). On the surface of the agar, *A. actinomycetemcomitans* forms small colonies, with a diameter of approximately 0.5–1.0 mm. Primary cultures are often difficult to lift off the agar surface. Typical colonies are star-shaped or shaped like crossed cigars, with irregular edges. In broth culture, the growth shows small particle-like opacity and often sticks to the flask wall. However, some strains grow into a homogeneously turbid culture after repeated cultures.

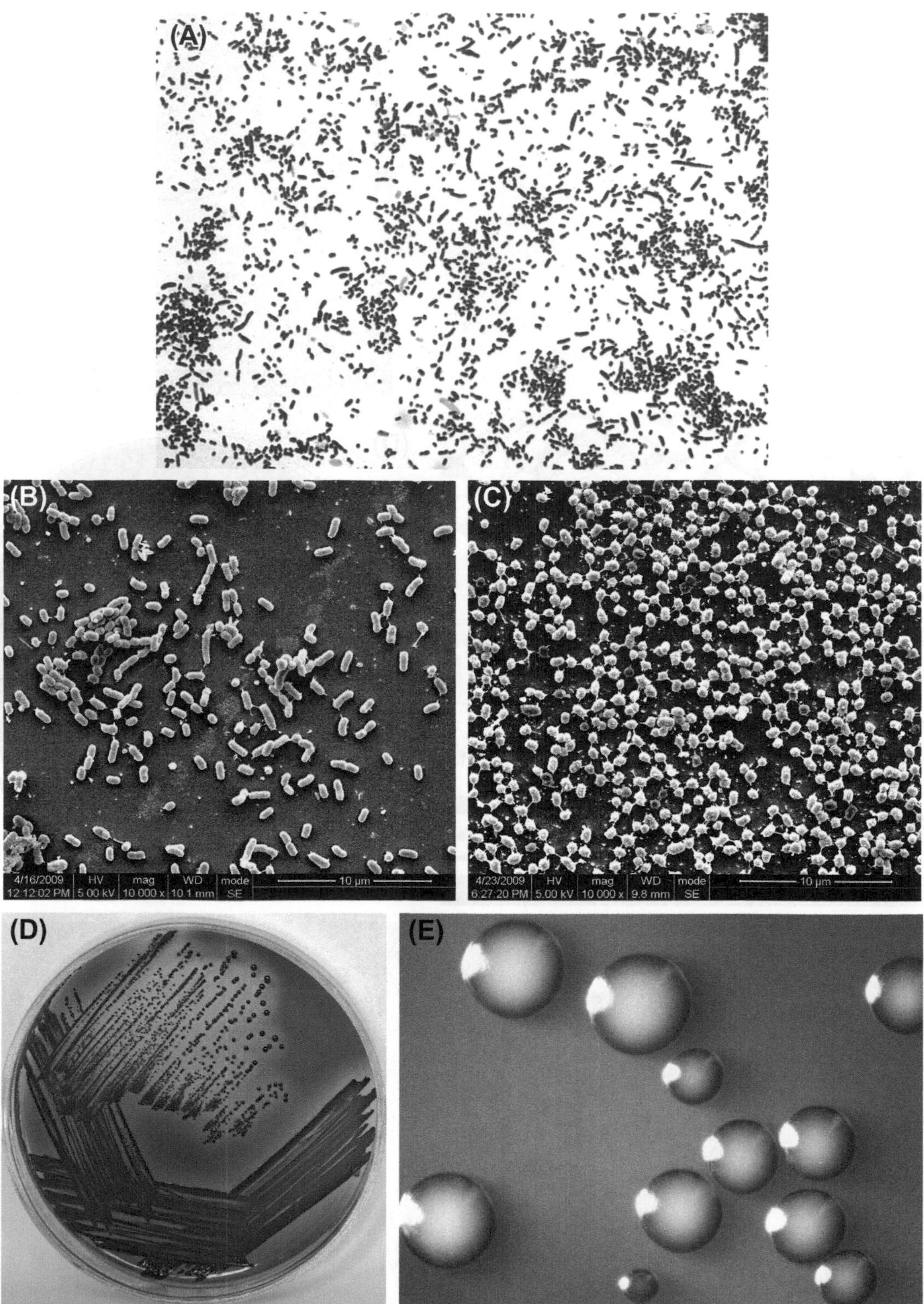

FIGURE 4.15 (A) *Prevotella intermedia* cells (Gram stain). (B) *P. intermedia* cells (SEM). Most *P. intermedia* cells form short rods and measure 0.4–0.7 µm × 1.5–2 µm, while some cells can measure up to 12 µm in diameter. Cells stain gram-negative. (C) *P. intermedia* cells (SEM). (D) *P. intermedia* colonies (BHI blood agar). (E) *P. intermedia* colonies (stereomicroscope). Most strains of *P. intermedia* form colonies 0.52.0 mm in diameter. Colonies are round, low, convex, translucent, with a smooth surface. Colonies show hemolytic reaction on the surface of blood agar, while aging or large colonies may appear opaque. After incubation for 48 h under anaerobic conditions, colonies may appear gray, brown, or black. Colonies can produce a hemolytic ring on rabbit blood agar. One-third of *P. intermedia* strains can produce dark brown to black colonies within 2 days. Colonies fluoresce brick red when exposed to long-wavelength UV light.

head and neck, and pleural infection. *P. intermedia* is occasionally isolated from blood, abdominal, or pelvic specimens.

The genomic GC content is 41–44%. The type strain is ATCC25611.

4.2.7.2 Prevotella nigrescens

Originally, this species was classified as a strain of *P. intermedia*. However, since it does not produce lipase, it can be thus distinguished from *P. intermedia*.

P. nigrescens is a gram-negative, nonsporulating, melanin-producing, obligate anaerobe. Cell morphologies are shown in Figure 4.16(A) and (B).

For culture requirements, refer to the information on *P. intermedia*. Characteristic colonies are shown in Figure 4.16(C) and (D)

Most strains ferment glucose, maltose, sucrose, and maltodextrin to produce acid. They also produce indole, hydrolyze starch, and hydrolyze gelatin. In liquid medium containing glucose, these bacteria can produce acetic acid, succinic acid, isobutyric acid, and isovaleric acid.

The genomic GC content is 40–44%. The type strain is ATCC33563 (namely NCTC9336, VPI8944).

4.2.7.3 Prevotella melaninogenica

P. melaninogenica is gram-negative, nonsporulating, melanin-producing obligate anaerobic bacteria. Bacterial cells are shown in Figure 4.17(A) and (B).

Most strains require hemin (1 mg/L) and vitamin K (0.1 mg/L) to grow and can grow at pH 8.5 and 25 °C. Characteristic colonies are shown in Figure 4.17(C), (D), and (E).

Usually, glucose broth culture turns turbid and is accompanied by smooth or filamentous precipitation. The final pH value range is from pH 4.6 to 5.0.

Clinical specimens are isolated from the gingival sulcus.

The genomic DNA GC content is 36–40%. The type strain is ATCC25845.

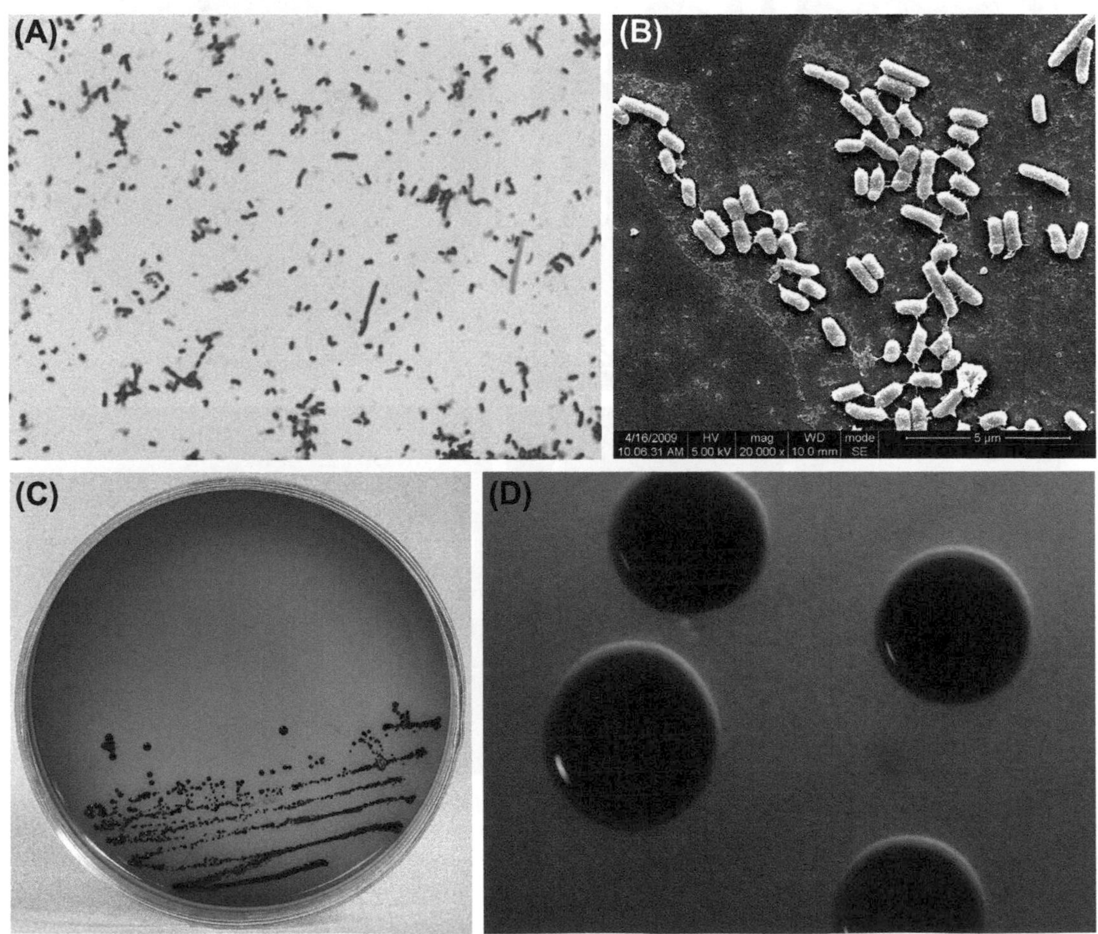

FIGURE 4.16 (A) *Prevotella nigrescens* cells (Gram stain). (B) *P. nigrescens* cells (SEM). Most cells of *P. nigrescens* are coccobacillus. In broth culture the cells are 0.3–0.4 µm × 1–2 µm in size. Some cells can measure up to 6–10 µm. Cells stain gram-negative. (C) *P. nigrescens* colonies (BHI blood agar). (D) *P. nigrescens* colonies (stereomicroscope). *P. nigrescens* on horse blood agar after 72 h. The colony diameter is 0.5–2 mm and appears circular, with a neat edge, low convex, smooth, and produces brown or black pigment. The edge of the colony is usually black, the center is cream-colored to dark brown. Most strains produce weak hemolytic reaction. Few strains produce a ring indicative of α-hemolysis.

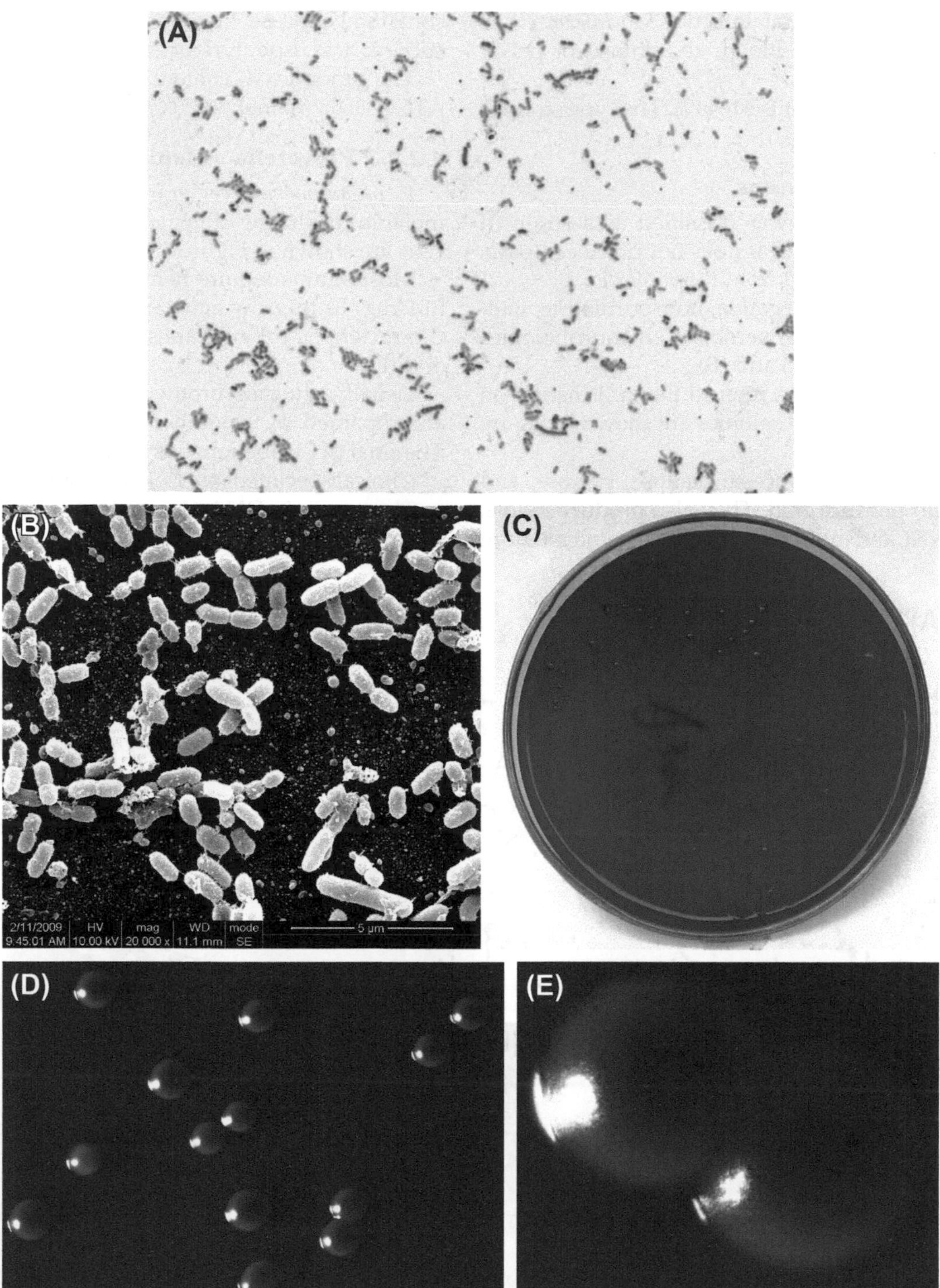

FIGURE 4.17 (A) *Prevotella melaninogenica* cells (Gram stain). (B) *P. melaninogenica* cells (SEM). *P. melaninogenica* cells are coccobacilli and measure 0.5–0.8 μm × 0.9–2.5 μm. Occasionally, cells longer than 10 μm are observed. Cells stain gram-negative. (C) *P. melaninogenica* colonies (BHI blood agar). (D) *P. melaninogenica* colonies (stereomicroscope). (E) *P. melaninogenica* colonies (stereomicroscope). *P. melaninogenica* colonies on blood agar are 0.5–2.0 mm in diameter. Colonies appear round, convex, glossy, with a neat edge. The centers of colonies are usually black and the edges are gray to light brown. After 5–14 days' culture, colonies become completely black. Few strains produce β-hemolytic reaction on rabbit blood agar.

4.2.7.4 Prevotella corporis

P. corporis is a gram-negative nonsporulating, melanin-producing obligate anaerobe. Examples of cells are shown in Figure 4.18(A) and (B).

Growth of *P. corporis* requires chlorinated hemoglobin and vitamin K1. A 10% final concentration of serum can promote growth and improve fermentation in some strains. Characteristic colonies are shown in Figure 4.18C

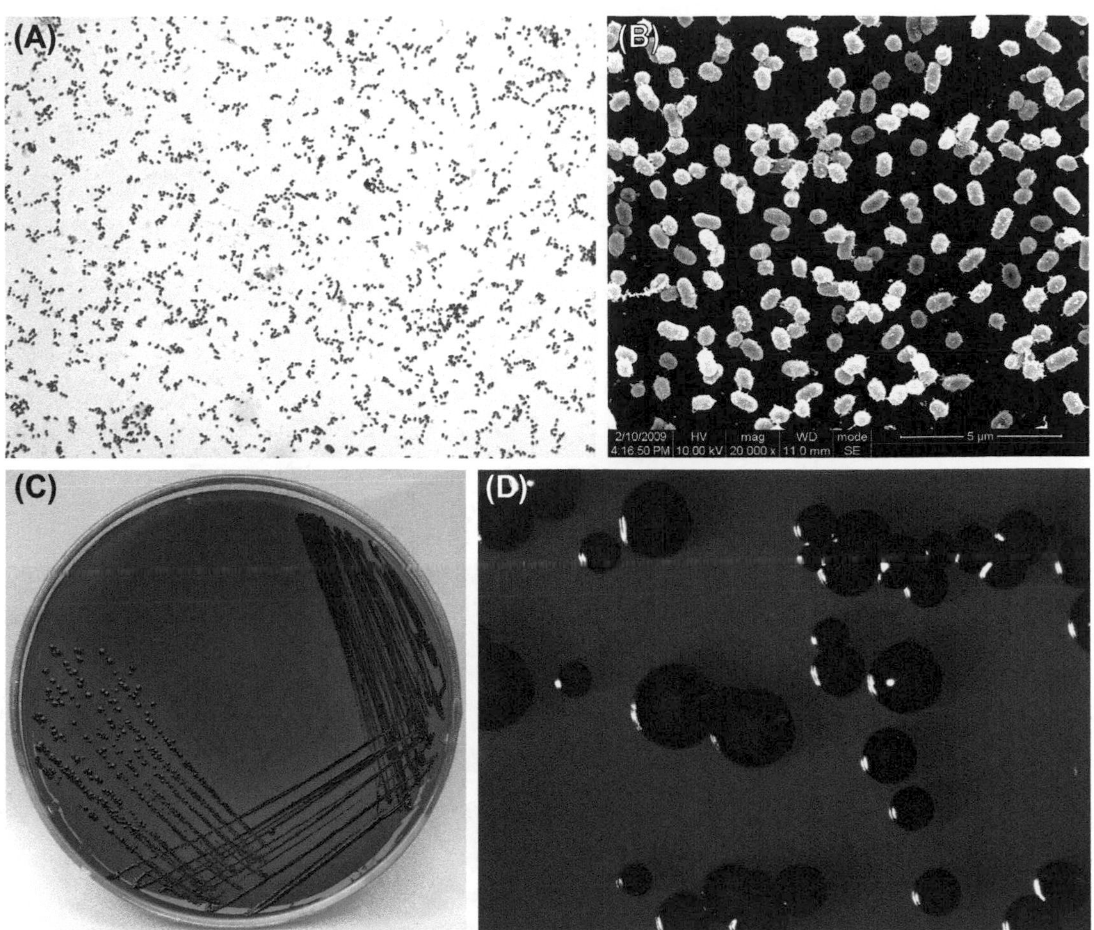

FIGURE 4.18 (A) *Prevotella corporis* cells (Gram stain). (B) *P. corporis* cells (SEM). *P. corporis* cells grown in glucose broth culture are measured 0.9–1.6 μm × 1.6–4.0 μm in size. Cells arrange singly, in pairs, or in short chains. Coccoid cells are also commonly seen, and long filamentous cells can sometimes be observed. Cells stain gram-negative. (C) *P. corporis* colonies (BHI blood agar). (D) *P. corporis* colonies (stereomicroscope). *P. corporis* cultured under anaerobic conditions on blood agar form colonies ranging from pinprick-sized to 1.0 mm. Colonies are rounded bumps with neat edges. After 48–72 h incubation, colonies can turn light yellow with brownish edges, and colonies cultured for 4–7 d turn dark brown.

and 4.18D. Glucose broth cultures turn cloudy and often have smooth or coarse precipitates adhered to the bottom flask. The final pH value of glucose broth cultures is 4.8–5.1.

P. corporis can be detected in specimens of all types of clinical infection, including the oral cavity.

The genomic GC content is 43–46%. The type strain is ATCC33457.

4.2.7.5 Prevotella loescheii

P. loescheii is a species of gram-negative, nonsporulating, melanin-producing, obligate anaerobic bacteria. Typical cells are shown in Figure 4.19(A), (B), and (C).

Cultures of *P. loescheii* require chlorinated hemoglobin, while the addition of 10% serum enhances fermentation. Colonies observed by stereomicroscope are shown in Figure 4.19(D) and (E).

Glucose broth cultures turn cloudy and are accompanied by a smooth deposit. The final pH value of cultures grown in glucose broth is between pH 4.9 and 5.4. *P. loescheii* does not produce H_2S in SIM medium. But the hydrolysis of esculin and the ability to ferment cellobiose help distinguish this species from *P. melaninogenica* and *P. dentocola*.

The genomic GC content is 46% (type strain). The type strain is ATCC15930 (NCTC11321).

4.2.8 *Porphyromonas*

In 1998, three species of *Bacteroides* were found to have significantly different biological characteristics than other *Bacteroides* species. However, they were similar in their ability to ferment carbohydrates to produce melanin. These species, namely *Bacteroides asaccharolyticus*, *Bacteroides gingivalis*, and *Bacteroides endodentalis* have been placed into a new genus: *Porphyromonas*. Members of this genus most commonly found in the oral cavity are *Porphyromonas gingivalis* and *Porphyromonas endodentalis*.

Broth cultured cells are typically small rods approximately 0.5–0.8 × 1.0–3.5 μm in size, but occasionally cells can be found that measure 4–6 μm long. These bacteria produce no spores, are nonmotile, and are gram-negative.

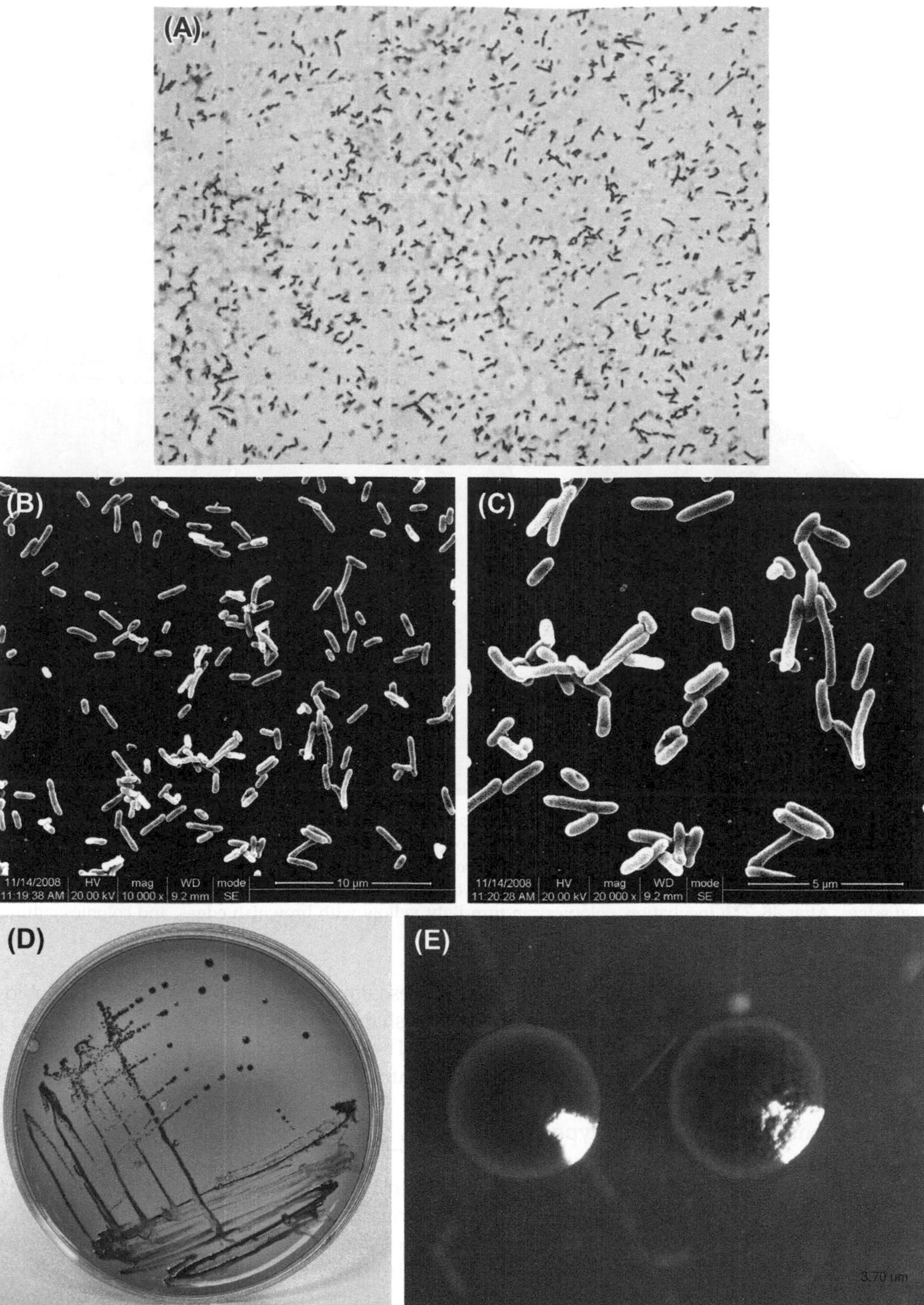

FIGURE 4.19 (A) *Prevotella loescheii* cells (Gram stain). (B) *P. loescheii* cells (SEM). (C) *P. loescheii* cells (SEM). Cells of *P. loescheii* grown in glucose broth culture are 0.4–0.6 µm × 0.8–15 µm in size. They are mostly coccobacilli, but can also be club-shaped and long rods. Cells organize into single cells, in pairs, or in a chain-like arrangement. Cells stain gram-negative. (D) *P. loescheii* colonies (BHI blood agar). (E) *P. loescheii* colonies (stereomicroscope). *P. loescheii* cultures grown on blood agar under anaerobic conditions form colonies 1.0–2.0 mm in diameter. Colonies are rounded, low convex, lustrous, translucent, with a smooth surface and neat edges. On agar plates containing whole blood, colonies turn white or yellow after 48 h. When anaerobic culture is continued for 14 days, colonies turn light brown. Some strains do not produce obvious dark brown or black colonies.

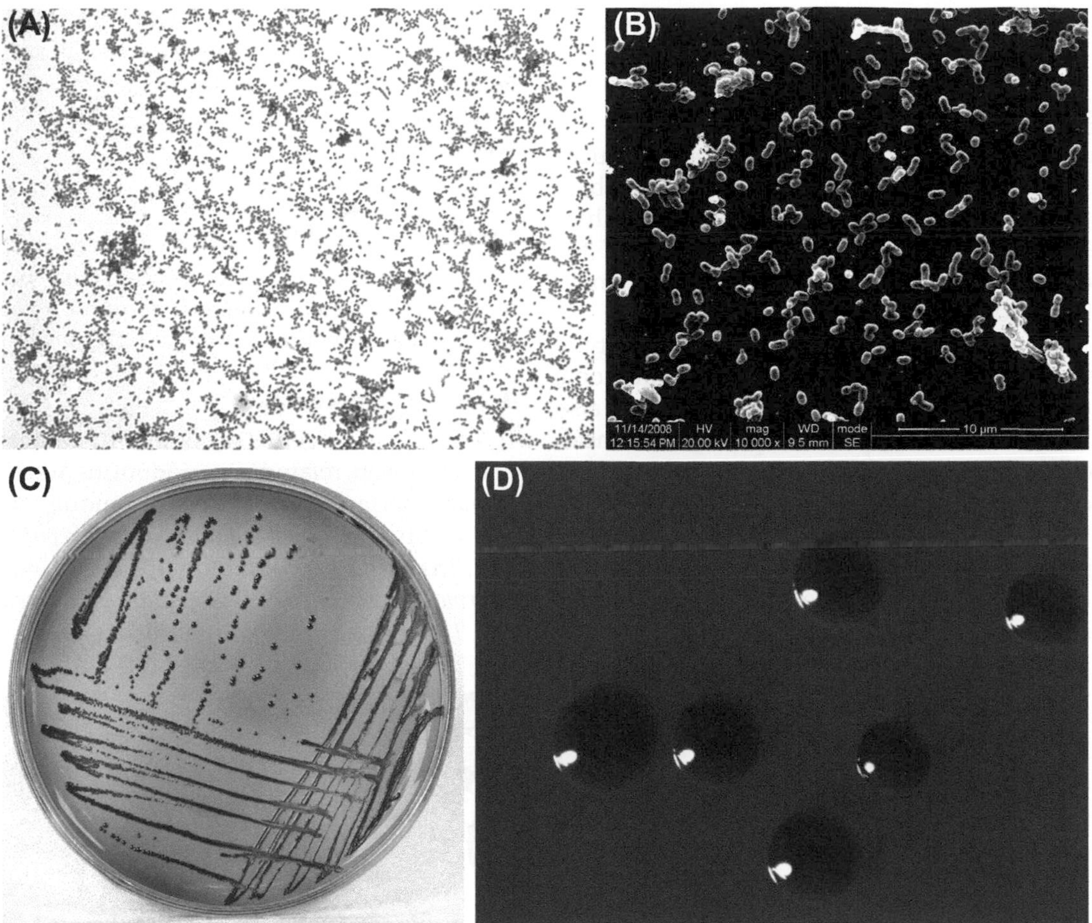

FIGURE 4.20 (A) *Porphyromonas gingivalis* cells (Gram stain). (B) *P. gingivalis* cells (SEM). *P. gingivalis* cells are 0.5 µm × 1–2 µm rods or coccobacilli, cells on solid medium form coccobacilli or very short rods. Cells stain gram-negative. (C) *P. gingivalis* colonies (BHI blood agar). (D) *P. gingivalis* colonies (stereomicroscope). *P. gingivalis* forms colonies 1–2 mm in diameter on blood agar. Colonies are rounded, lustrous, with a smooth (or occasionally rough) surface. After 4–8 days' culture, melanin spreads from the edge to the center of the colony to form black colonies. A small number of colonies do not produce melanin.

Porphyromonas are obligate anaerobes with an optimum growth temperature of 37 °C. On blood agar plates, they can form colonies 1–3 mm in diameter that are protuberant, lustrous, and with smooth surface (very few with rough surface).

The primary endpoint product in PYG medium is butyric acid and acetic acid. There is a small amount of propionic acid, isobutyric acid, and isoamyl propionate produced.

The genomic GC content of this genus is 46–54%. The type species is *Porphyromonas asaccharolytica*.

4.2.8.1 Porphyromonas gingivalis

P. gingivalis are gram-negative, obligate anaerobic bacteria that do not form spores and produce melanin, as shown in Figure 4.20(A) and (B). Its cell wall peptidoglycan contains lysine, while the main respiratory quinone has nine isoprene units of unsaturated methyl naphthalene quinones.

As an obligate anaerobe, chlorinated hemoglobin and vitamin K1 are required in the growth media. Hemoglobin is the main product of porphyrin. Characteristic colonies are shown in Figure 4.20(C) and (D).

When grown in BM or PYG media, the main terminal acids produced by *P. gingivalis* are butyric acid and acetic acid, with a small amount of propionic acid, isobutyric acid, isoamyl propionate, and phenylacetic acid produced as well.

P. gingivalis produces indoles, does not produce alpha fucosidase, does not reduce nitrate to nitrite, and does not hydrolyze esculin and starch. Cells test positive for MDH and glutamate dehydrogenase, negative for glucose phosphate dehydrogenase, and glucose 6-phosphate acid salt dehydrogenase. These enzymatic characteristics, as well as its ability to produce acetic acid, are the important features that differentiate *P. gingivalis* from other morphologically similar bacteria.

Its DNA GC content is 46–48%. The type strain is ATCC33277.

4.2.8.2 Porphyromonas endodontalis

This gram-negative, obligate anaerobe produces melanin. Cells do not form spores. Examples of bacterial cells are shown in Figure 4.21(A) and (B). Chlorinated hemoglobin and vitamin K1 are required for culture growth. Characteristic colonies are shown in Figure 4.21(C) and (D).

The primary terminal acid products are N-butyric acid and acetic acid, but a small amount of propionic acid, isobutyric acid, and isoamyl propionate are also produced. The cells do not produce phenylacetic acid. *P. endodontalis* can hydrolyze gelatin and arginine. Its protein hydrolysis activity is extremely low. It does not hydrolyze esculin and starch and can produce indoles and hydrogen sulfide. It does not produce alpha fucosidase and does not reduce nitrate to nitrite. Cells test negative for sheep cell agglutination (SCAT) but are positive for MDH and glutamate dehydrogenase. *P. endodontalis* is negative for glucose phosphate dehydrogenase and glucose 6-phosphate dehydrogenase.

The genomic GC content is 49–51%. The type strain is ATCC 35406.

4.2.9 *Treponema*

Spirochetes are a type of slender, curved spiral, highly motile gram-negative bacteria. Common genera of spirochetes are: *Spirochaeta*, *Cristispira*, *Treponema*, and *Borrelia*.

4.2.9.1 Treponema

Treponema is a genus of commonly found oral bacteria that are closely related to periodontitis and the etiology of implant periarthritis. Species commonly detected in the oral cavity are *Treponema denticola*, *Treponema scaliodontum*, *Treponema macrodentium*, *Treponema oralis*, *Treponema intermedia*, *Treponema maltophilum*, *Treponema socranskii*, and *Treponema vincentii*. In addition, the gastrointestinal

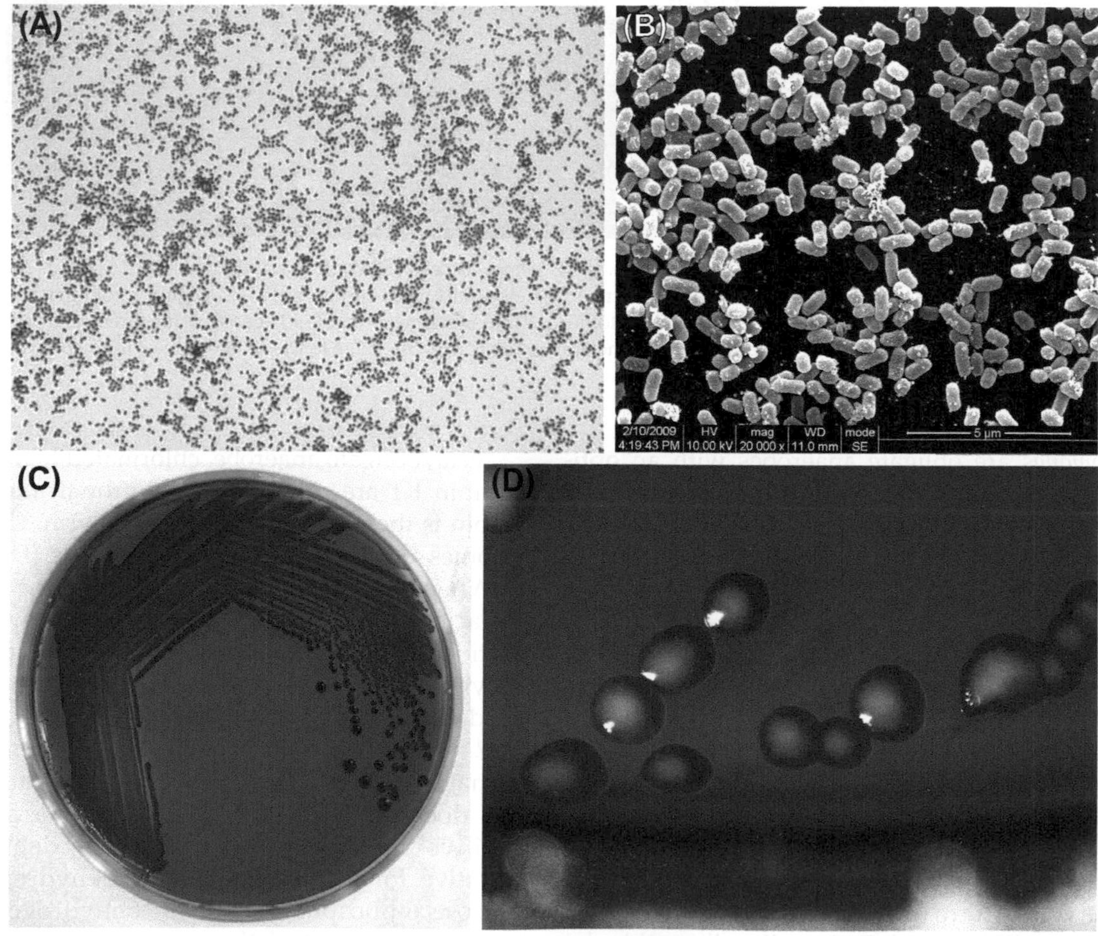

FIGURE 4.21 (A) *Porphyromonas endodontalis* cells (Gram stain). (B) *P. endodontalis* cells (SEM). *P. endodontalis* organizes as single cells approximately 0.4–0.6 µm × 1.0–2.0 µm in size. They do not form spores, are nonmotile, and are gram-negative. (C) *P. endodontalis* colonies (BHI blood agar). (D) *P. endodontalis* colonies (stereomicroscope). *P. endodontalis* on blood agar forms rounded, lustrous colonies with a smooth surface. The colonies have a neat edge and gain progressively deeper dark brown or black pigmentation (hemoglobin is the main product of porphyrin) after culturing for 47 days. Most strains form solid adhesion to the surface of blood agar but are slow growing in liquid medium.

tract and the vagina are also major colonization sites for bacteria of this genus in humans.

Clinical samples of *Treponema* are ideally observed with a dark field or a phase contrast microscope. *Treponema* cells are gram-negative, but most of the strains do not take up stain easily by Gram staining or Giemsa staining. Silver impregnation stain and Ryu's stain are better for the observation of *Treponema* cells. Presently, we commonly use the Congo red negative stain method, as it is not only economic and simple but the helical cells are also very easy to observe (Figure 4.22(A)). The morphology of the bacterial cells can be seen under SEM (Figure 4.22(B) and (C)).

The outer membrane of *Treponema* cells is similar to the outer membrane of gram-negative bacteria cells. The content includes lipids, proteins, and carbohydrates. The lipids are mainly made up of phospholipids and glycolipids, while the cell wall contains muramic acid, glucosamine, and ornithine. Peptidoglycan accounts for 10% of the dry weight of the cell.

The genus *Treponema* consists of obligate anaerobes, but those species that are pathogenic to humans may be microaerophiles. *Treponema* are heterotrophic bacteria that mainly metabolize through fermentation. They can use a variety of carbohydrates and amino acid as their carbon source and energy. *Treponema* species are difficult to grow in artificial culture media, the growth of some species requires long chain fatty acids from serum, and other species require branched-chain fatty acids. *Treponema denticola*, *Treponema vincentii*, and *Treponema scaliodontum* require cocarboxylase in serum.

The genomic GC content of most *Treponema* species ranges from 25% to 54% (by Tm). The type species is *T. pallidum*.

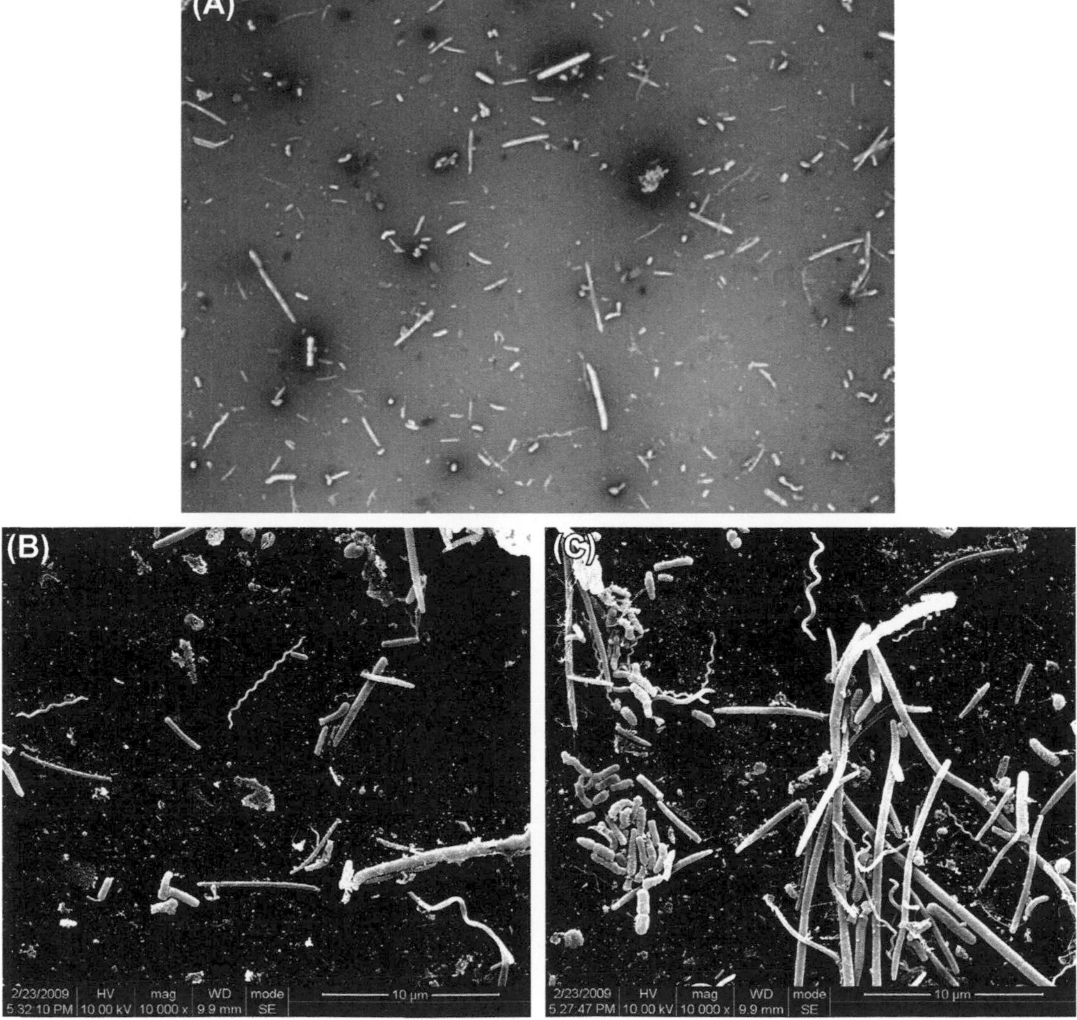

FIGURE 4.22 (A) Plaque samples of *Treponema* cells (Congo red negative staining). (B) Plaque samples of large, medium, and small *Treponema* cells (SEM). (C) Plaque samples of *Treponema* and *Borrelia* cells (SEM). *Treponema* cells measure 0.1–0.4 μm × 5–20 μm. They form a spiral rod and can be either a tight, regular, or irregular spiral. Cells have one or more axial flagella inserted into the plasma membrane at opposite ends of the rod. In old cultures, huge bubbles and spiral spheres can be observed. Large, medium, and small spirochetes can be observed in clinical samples. The spiral-shaped cells measure 0.2–0.5 μm × 3–20 μm, and form regular or irregular loose spirals. The cells stain gram-negative.

CHAPTER

5

Oral Mucosal Microbes

5.1 GRAM-POSITIVE BACTERIA

5.1.1 *Staphylococcus*

5.1.1.1 Staphylococcus aureus

S. aureus is a gram-positive species of bacteria. Its peptidoglycan structure is L-Lys-Gly5-6 and the cell wall teichoic acid is formed by ribitol polymerization. Cellular morphology is shown in Figure 5.1(A) and (B). The genomic GC content is 32–36% and the type strains are ATCC126000, NCTC8352, and CCM885.

S. aureus is a facultative anaerobe but also grows well under aerobic conditions (Figure 5.1(C), (D), and (E)). After incubation on semisolid thioglycollate medium, colonies grow rapidly and are uniform and dense. The final pH of glucose broth incubated with *S. aureus* is always between pH 4.3 and 4.6.

S. aureus is associated with human infections, such as facial furuncle and carbuncle, loose connective tissue inflammation, and tumor postoperative wound infections.

S. aureus is able to produce golden pigments on common agar plate. The colonies are round, raised, shiny opaque, and have a smooth surface and a whole edge. The diameter is 2–3 mm. A β-transparent hemolytic zone surrounds the colony when *S. aureus* is incubated on blood BHI agar.

5.1.2 *Streptococcus*

5.1.2.1 Streptococcus mitis

S. mitis cells are gram-positive and spherical or elliptical in shape (about 0.6–0.8 μm in diameter). They can form long chains in broth culture (Figure 5.2(A)). The GC content of the *S. mitis* genome is 38–39%, and its type strain is NCTC3165.

Streptococcus mitis cells grow variably at 45 °C, and the final pH value of a culture in glucose broth is around pH 4.2–5.8. Cells in broth culture change from rough to smooth following subculture, and cells that are grown under aerobic conditions can be smooth or rough. TPY agar is the common medium on which *S. mitis* is grown (Figure 5.2(B)). Significant α-hemolytic reaction can be observed when *S. mitis* is grown on blood agar plates (Figure 5.2(C)). *S. mitis* forms small broken-glass-like colonies on sucrose agar plates, and very few strains can form the typical sticky colonies. Colonies observed by stereomicroscope are shown in Figure 5.2(D).

Biochemical reactions. *S. mitis* can ferment glucose, maltose, sucrose, lactose, and salicin to produce acid, and some strains can also ferment raffinose and trehalose to produce acid. However, inulin, mannitol, sorbitol, glycerol, arabinose, and xylose cannot be fermented by *S. mitis*.

Colonization characteristics. *S. mitis* can be detected in human saliva, oral mucosa, sputum, or excrement. It is a common oral *Streptococcus* and a member of the oral microflora.

5.1.2.2 Streptococcus pyogenes

S. pyogenes belongs to Lancefield group A and is gram-positive (Figure 5.3(A)). Cells of this species have a diameter of 0.5–1.0 μm and are usually spherical. However, cells from old cultures may be oval. Moreover, *S. pyogenes* cells are often arranged in short, medium, or long chains, but most cells in broth culture form long chains. Cells of clinical isolates are usually arranged in pairs. *S. pyogenes* cells observed by stereomicroscope are shown in Figure 5.3(B). The GC content of the *S. pyogenes* genome is 34.5–38.5% (Tm method), and its type strain is ATCC12344.

S. pyogenes is a facultative anaerobe, and the optimal temperature for growth is 37 °C. A nutrient-rich medium is required for *S. pyogenes* growth, and both blood and serum can promote its growth. Colonies growing on MS agar are shown in Figure 5.3(C). Cells from overnight blood agar plate culture have three different colony types: (1) colonies with β-hemolytic zone; (2) mucoid colonies; (3) lackluster colonies. In fact,

Atlas of Oral Microbiology
http://dx.doi.org/10.1016/B978-0-12-802234-4.00005-7

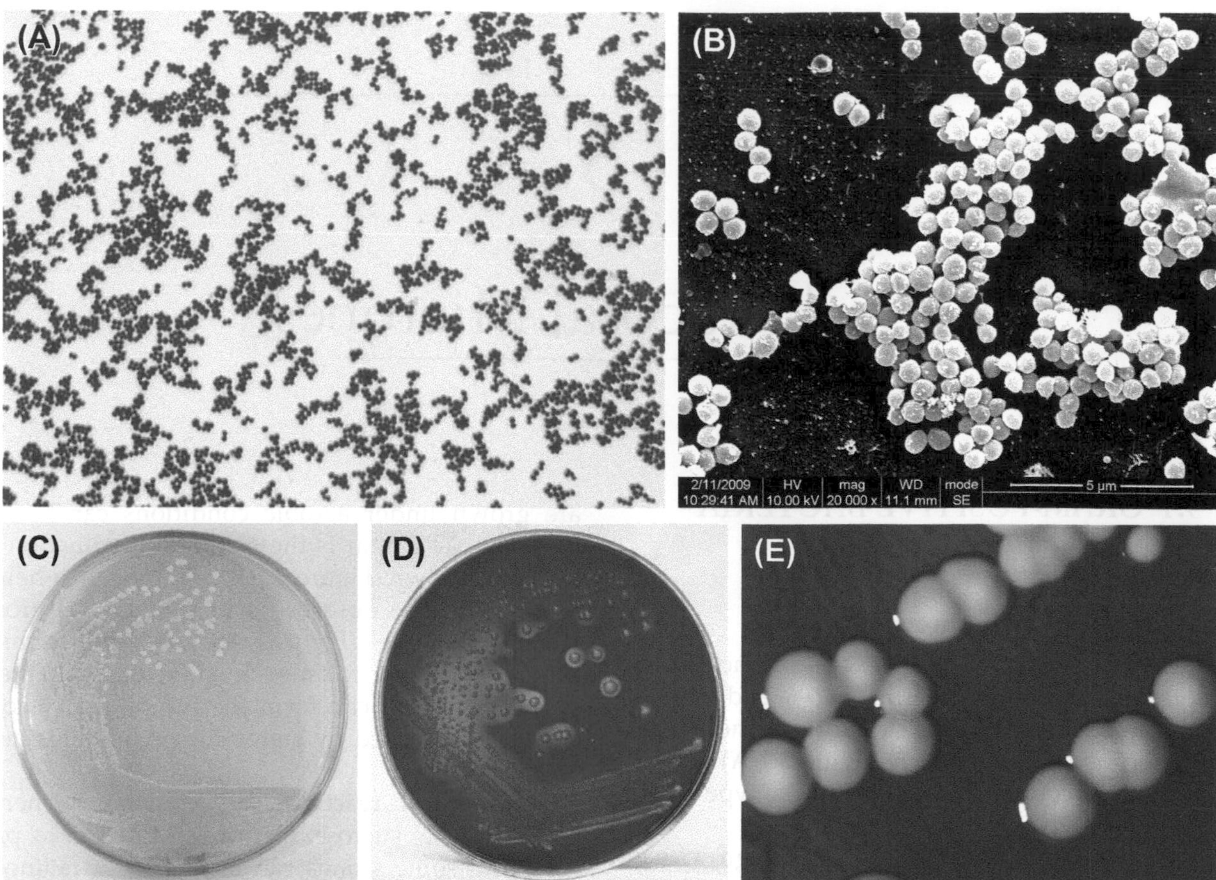

FIGURE 5.1 (A) *Staphylococcus aureus* cells (Gram stain). (B) *S. aureus* cells (SEM). The cells of *S. aureus* are characterized as being spherical (0.5–1.5 μm in diameter), gram-positive cocci arranged as single cells, pairs, tetrads, and clusters. (C) Golden pigment of *S. aureus* incubated on common agar plate. (D) β-hemolytic zone of *S. aureus* incubated on blood BHI agar plate. (E) Colonies of *S. aureus* (stereomicroscope).

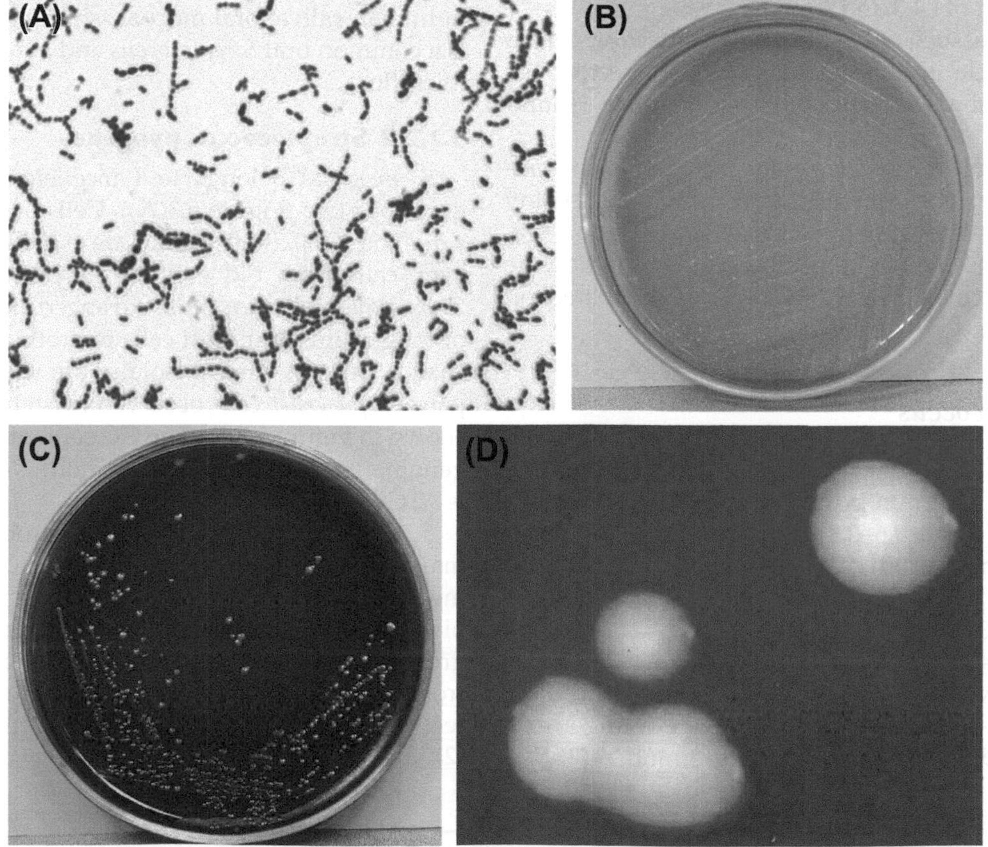

FIGURE 5.2 (A) Cells of *S. mitis* (Gram stain). (B) Colonies of *S. mitis* (TPY agar plate). (C) Colonies of *S. mitis* (BHI agar plate). (D) Colonies of *S. mitis* (stereomicroscope).

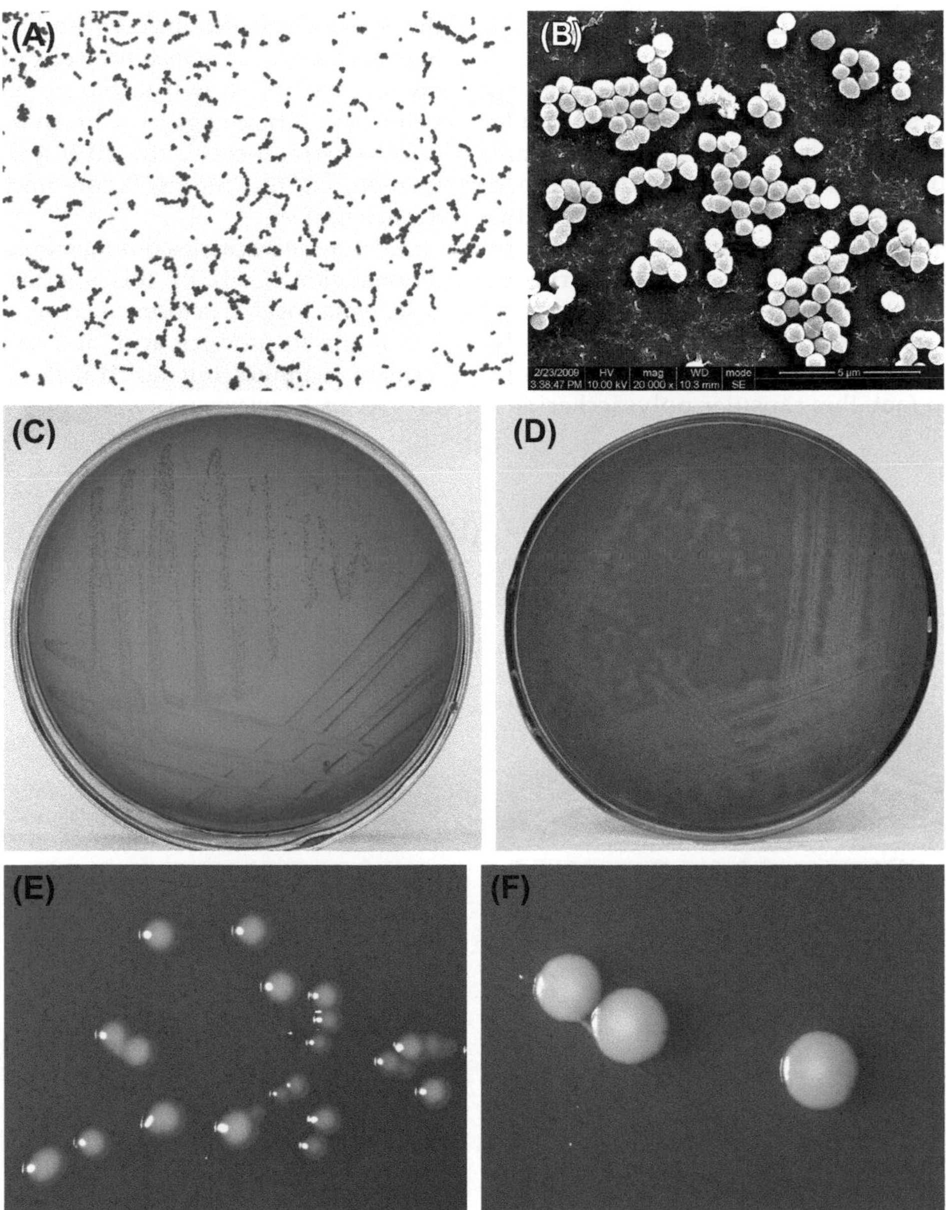

FIGURE 5.3 (A) Cells of *S. pyogenes* (Gram stain). (B) Cells of *S. pyogenes* (SEM). (C) Colonies of *S. pyogenes* (MS agar plate). (D) Colonies of *S. pyogenes* with β-hemolytic zone (BHI-blood agar plate). (E) Mucoid colonies of *S. pyogenes* (BHI-blood agar plate, stereomicroscope). (F) Lackluster colonies of *S. pyogenes* (BHI-blood agar plate, stereomicroscope).

the type of colony is mainly related to the growth condition and production of hyaluronidases.

S. pyogenes can produce several pathogenic virulence factors, including bacteriocins, hemolysin O, streptokinase, NADH enzyme, hyaluronidase, and M protein. The main sites of colonization for *S. pyogenes* are the dental plaque, hypopharynx, and the upper respiratory tract. Clinical samples can be isolated from skin lesions, inflammatory secretions, or blood. *S. pyogenes* can also be found in loose connective tissue inflammation in the maxillofacial region, pulpitis, or infection after exelcymosis.

5.1.2.3 Streptococcus pneumoniae

S. pneumoniae cells are spherical or elliptic (about 0.5–1.25 μm in diameter) and mostly arranged in pairs. They can occasionally be found in short chains or as single cells. Typical cells of this species are lanceted and arranged in pairs, and the extracellular capsule can be observed by capsule staining. However, cells can easily form chains after continuous culture. Cells of *S. pneumoniae* are gram-positive but can change to be gram-negative when aged (Figure 5.4(A) and (B)). *S. pneumoniae* cells observed by SEM are shown in Figure 5.4(C). The GC content of the *S. pneumoniae* genome is 30%

(by Tm method) or 42% (by Bd method), and its type strain is NCTC7465.

S. pneumoniae is a kind of facultative anaerobe, but could generate a substantial amount of H_2O_2 under aerobic conditions. Growth of *S. pneumoniae* requires a complex medium rich in nutrients, including blood, serum, or ascites, as well as vitamin B. Cells of this species grow well on BHI-blood agar plates, and form roof-like, reflective and smooth colonies or rugose, mycelium-like, rough colonies (Figure 5.4(D)). Cells of *S. pneumoniae* can generate a great deal of capsular polysaccharides and therefore often form sticky colonies (Figure 5.4(E)). In fact, this capsular polysaccharide is the species-specific antigen and virulence factor of *S. pneumoniae*. *S. pneumoniae* cells observed by stereomicroscope are shown in Figure 5.4(F). The final pH value of a culture in glucose broth is pH 5.0. Compared to other streptococci, the addition of bile or cholate is required for the isolation and identification of *S. pneumoniae*.

S. pneumoniae carries out metabolic reactions through fermentation. Cells of this species can ferment glucose, galactose, fructose, sucrose, maltose, raffinose, and inulin to produce acid, and some strains can also ferment mannitol but not dulcite and sorbitol. Bacteriolysis of *S. pneumoniae* cells by bile is positive, and

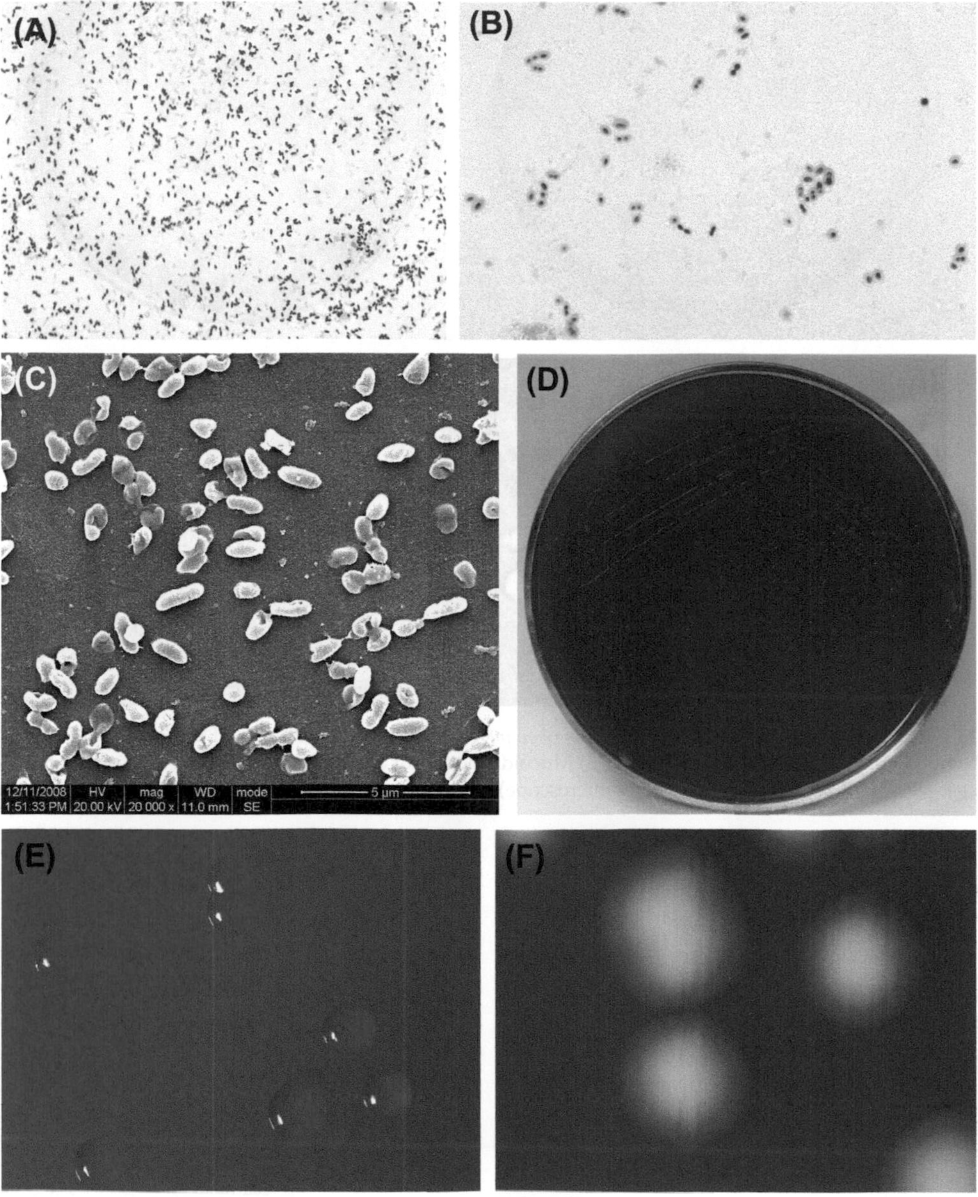

FIGURE 5.4 (A) Lanceted cells of *S. pneumoniae* (Gram stain). (B) Capsule of *S. pneumoniae* cells (capsule stain). (C) Cells of *S. pneumoniae* (SEM). (D) Colonies of *S. pneumoniae* (BHI-blood agar plate). (E) Sticky colonies of *S. pneumoniae* (BHI-blood agar plate, stereomicroscope). (F) Colonies of *S. pneumoniae* (stereomicroscope).

this characteristic is helpful in distinguishing *S. pneumoniae* from other streptococci.

S. pneumoniae mainly inhabits the upper respiratory tracts of healthy individuals or livestock and can be isolated from clinical samples of amygdalitis, pneumonia, meningitis, and otitis media. However, the colonization, distribution, and pathogenicity of this species in oral cavity are still unknown.

5.1.2.4 Streptococcus vestibularis

S. vestibularis gets its name because this species was first isolated from the vestibule of the human oral cavity. The cells of *S. vestibularis* are gram-positive, spherical (about 1 μm in diameter), and arranged in chains (Figure 5.5(A)). *S. vestibularis* cells observed by SEM are shown in Figure 5.5(B). The GC content of the *S. vestibularis* genome is 38–40%, and its type strain is TC12166 (= MM1).

S. vestibularis is a facultative anaerobe, and the optimal temperature for growth is 37C, but cells of this species can also still grow at 10–45C. Except the type strain MM, the majority of strains belonging to this species can grow in media containing 10% bile. However, the species cannot grow in medium containing 4% NaCl, 0.004% crystal violet, or 40% bile. Alpha-hemolysis can be detected for all strains of this species growing on horse blood agar plates. In addition, black blue, lackluster, and umbilicate colonies (about 2–3 μm in diameter) with wavy edges can be observed when cells are grown on MS agar plates under anaerobic conditions (37C, 72 h culture), while black blue, glossy, and raised colonies (about 1–2 μm in diameter) with neat edges can be observed when cells are grown on MS agar plates under aerobic conditions (Figure 5.5(C)). Colonies observed by stereomicroscope are shown in Figure 5.5(D).

S. vestibularis can ferment N-acetylglucosamine, myricitrin, fructose, galactose, glucose, lactose, maltose, mannose, salicin, and sucrose to produce acid, but they do not ferment adonitol, arabinose, dextrin, dulcite, fucose, glycerol, glycogen, inositol, inulin, mannitol, melezitose, D-melibiose, raffinose, ribitol, sorbitol, starch, and xylose. The majority of strains of this species can ferment cellose and amygdalin to produce acid, and a minority of strains of this species can ferment trehalose and D-glucosamine to produce acid. *S. vestibularis* can also generate urease and H_2O_2 but not extracellular or intracellular polysaccharides. It also does not generate ammonia from arginine. In addition, the majority strains of this species can produce butanone with two hydroxyl groups and can hydrolyze esculin and starch.

S. vestibularis is mainly isolated from the mucosa or vestibule of the human oral cavity.

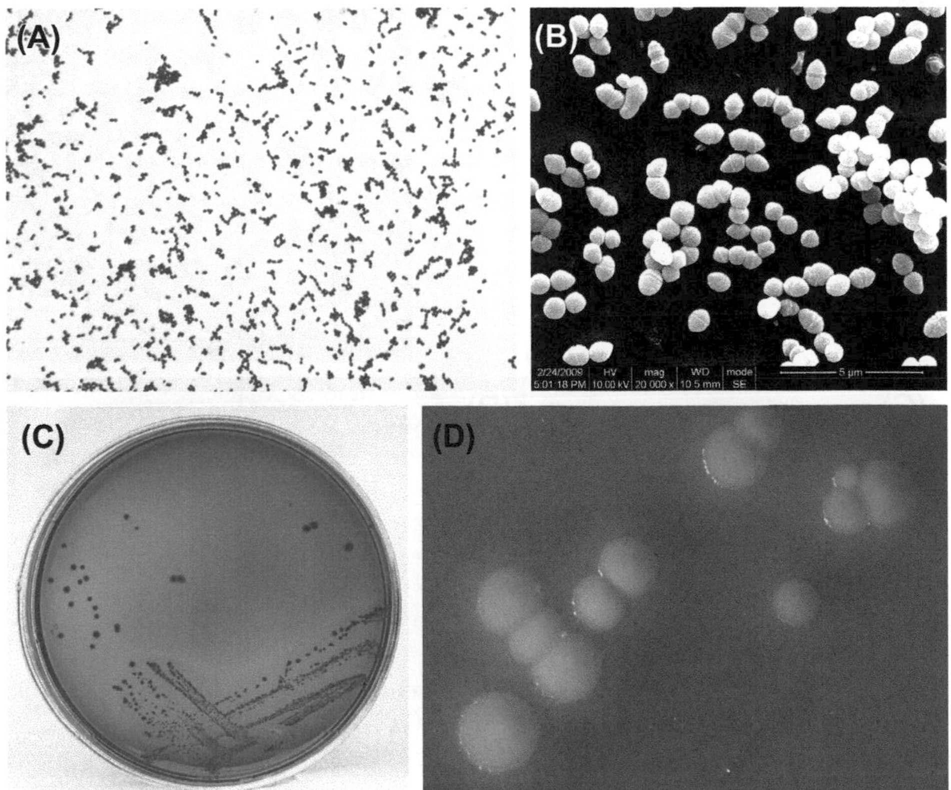

FIGURE 5.5 (A) Cells of *S. vestibularis* (gram-positive coccus). (B) Cells of *S. vestibularis* (SEM). (C) Colonies of *S. vestibularis* (MS agar plate). (D) Colonies of *S. vestibularis* (stereomicroscope).

5.2 GRAM-NEGATIVE BACTERIA

5.2.1 *Escherichia*

5.2.1.1 Escherichia coli

Commonly known as *E. coli* (Figure 5.6(A) and (B)), this is the most common bacteria belonging to the family *Enterobacteriaceae* found in the human body. *E. coli* is the first gram-negative bacillus that was detected in the neonatal oral cavity, and is thought to be transmitted from the mother during vaginal birth.

E. coli is a facultative anaerobe with low nutritional requirement, which can grow well under aerobic conditions. Characteristic colonies are shown in Figure 5.6(C) and (D).

E. coli occasionally can cause maxillofacial infections.

E. coli colonies are soft gray-white with a diameter of 2–3 mm.

5.2.2 *Haemophilus*

Haemophilus is a genus of gram-negative facultative anaerobic bacilli, named due to their requirement for blood during growth. Most bacteria of this genus are part of the normal microflora of the oral cavity and of the nasopharynx in human and animals. The main colonization site within the oral cavity is dental plaque, followed by saliva, and soft palate.

Species detected in the oral cavity include *H. actinomycetemcomitans*, *H. influenzae*, *H. parainfluenzae*, *H. aphrophilus*, *H. paraphrophilus*, *H. parahaemolyticus*, *H. paraprohaemolyticus*, and *H. segnis*. In the latest classification, *H. actinomycetemcomitans*, *H. aphrophilus*, and *H. segnis* were classified into a new genus: *Aggregatibacter*, and renamed *A. actinomycetemcomitans*, *A. aphrophilus* and *A. segnis*, respectively.

Generally, *Haemophilus* cells are club-shaped or rod-shaped and less than 1 μm in width. The length of cells can be short to medium. Sometimes cells are filamentous and can be pleiomorphic. Bacteria of this genus are gram-negative, do not produce spores, have no capsule, and are nonmotile.

Haemophilus are chemoorganotrophic bacteria. After 48 h culture on blood agar surface at 37 °C, most species will form flat or raised, colorless or pale-yellow, smooth colonies with a diameter of 0.5–2.0 mm. A few species like *H. parainfluenzae* form rough colonies. *H. parahaemolyticus* and *H. paraprohaemolyticus* can produce β-hemolytic reaction on blood agar. In broth containing glucose, the

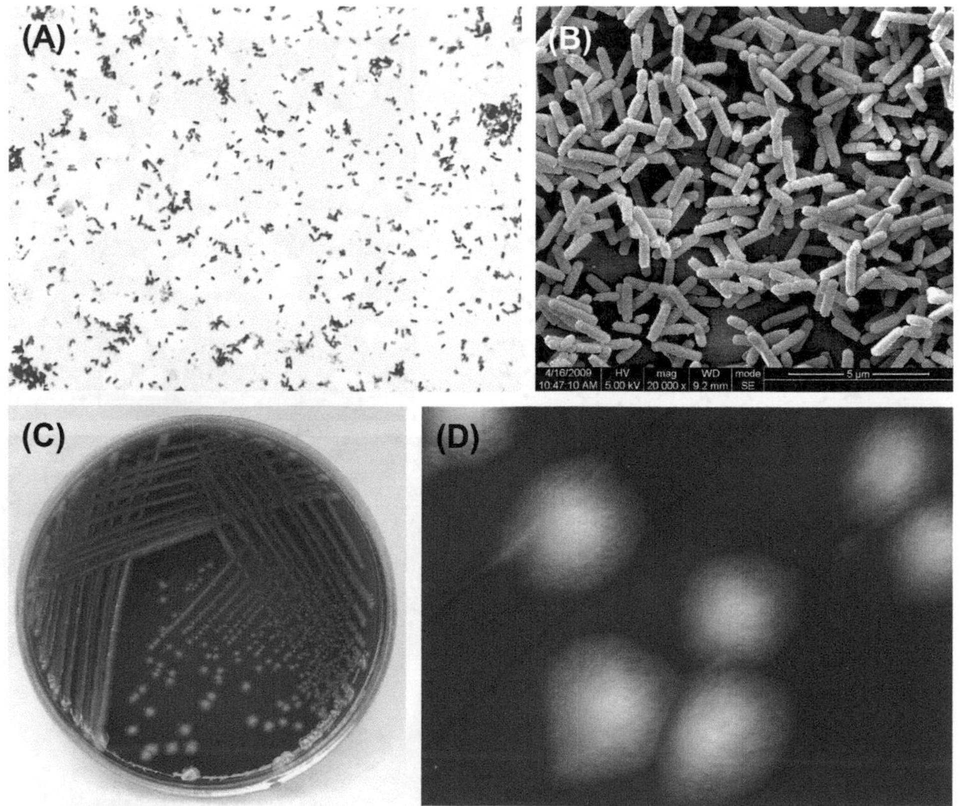

FIGURE 5.6 (A) *Escherichia coli* cells (Gram stain). (B) *E. coli* cells (SEM). The size of *E. coli* cell is 0.4–0.7 μm × 1–3 μm. Gram stain is negative. (C) *E. coli* colonies (BHI blood agar). (D) *E. coli* colonies (stereomicroscope).

terminal acid metabolites are acetic acid, lactic acid, and succinic acid.

Most oral strains of *Haemophilus* are nonpathogenic members of the normal human oral mucosa and upper respiratory tract. Occasionally, they can cause mixed infections such as periodontal abscess or jaw infection as conditional pathogens.

The genomic GC content ranges from 37% to 44% (by Tm method). The type species is *H. influenzae.*

5.2.2.1 Haemophilus influenzae

H. influenzae is a gram-negative bacillus. Cell morphology is shown in Figure 5.7(A) and (B).

H. influenzae is a facultative anaerobe. Atmosphere supplemented with 5–10% CO_2 is the most suitable growth environment. Cultures require growth factors from blood, especially factor X and factor V. Chocolate agar is the preferred growth medium. Characteristic colonies are shown in Figure 5.7(C), (D), and (E).

H. influenzae can ferment glucose and sucrose to produce acid, but fermentation of galactose, fructose, maltose, and xylose does not produce acid. They can also reduce nitrate but not nitrite. The species tests positive for catalase, while the *o*-nitrophenyl-β-D-galactopyranoside (ONPG) test shows up negative.

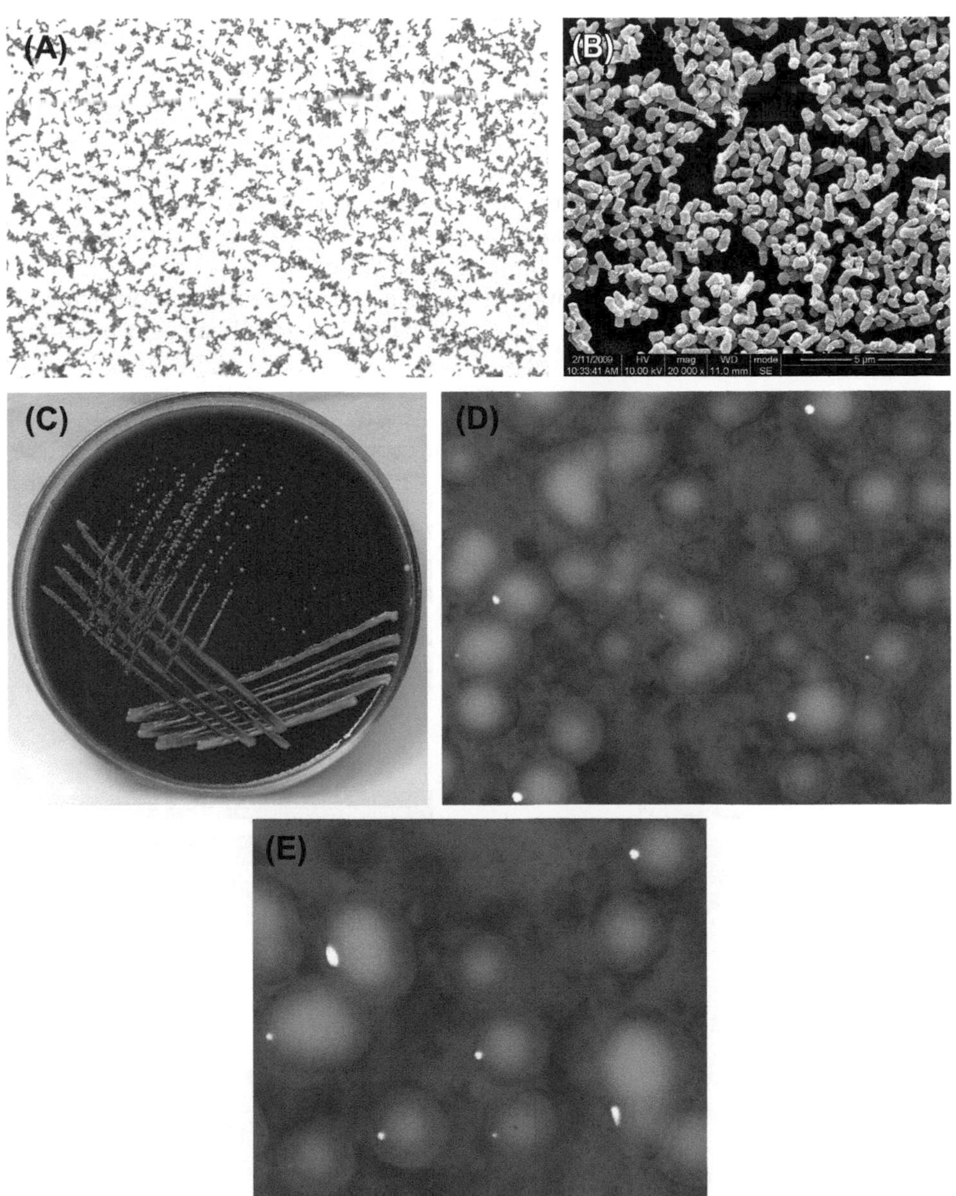

FIGURE 5.7 (A) *Haemophilus influenzae* cells (Gram stain). (B) *H. influenzae* cells (SEM). *H. influenzae* cells are gram-negative, club-shaped or small rods, with dimensions of 0.3–0.5 μm × 0.5–3.0 μm. The cells have no spores or capsule and are nonmotile. (C) *H. influenzae* colonies (BHI blood agar). (D) *H. influenzae* viscous colonies (stereomicroscope). (E) *H. influenzae* colonies (stereomicroscope).

H. influenzae can be detected in the nasopharynx of up to 75% of healthy children. However, the infection rate in adults is very low. The detection rate of capsule-producing strains in the healthy nasopharynx is only 3–7%.

The genomic GC content is 39% (by Tm method). The type strain is NCTC8143 (biological variant type II without capsule).

After cultivation on chocolate agar for 24 h, convex, gray, translucent smooth colonies form with a diameter of 0.5–1 mm. Strains with capsule usually form 1–3 mm large viscous colonies.

5.2.3 *Moraxella*

Moraxella, Neisseria, and *Kingella* all belong to the family *Neisseriaceae.*

5.2.3.1 Moraxella catarrhalis

M. catarrhalis are common bacteria belonging to the genus *Moraxella* found in the oral cavity. They were previously known as *Neisseria catarrhalis* and *Branhamella catarrhalis.*

The cells are gram-positive cocci and are shown in Figure 5.8(A) and (B).

M. catarrhalis are aerobic and can grow on agar medium. However, cultures grow better on blood agar. Characteristic colonies are shown on Figure 5.8(C) and (D).

The main site of colonization is the upper respiratory tract. This species generally does not cause disease but can cause catarrhal inflammation, such as acute pharyngitis and otitis media.

The genomic GC content is 40–43%. The type strain is ATCC25238 (NCTC11020).

After 48 h culture on blood agar, smooth, opaque, relatively flat, gray colonies form with a diameter of about 2 mm.

5.2.4 *Neisseria*

Neisseria are aerobic gram-negative diplococci belonging to the family *Neisseriaceae,* which mainly colonize the human oral cavity and nasopharynx. Most *Neisseria* are members of the normal microflora of the human body and are usually nonpathogenic. However, *N. meningitidis* and *N. gonorrhoeae* are important pathogens.

Common *Neisseria* in the oral cavity are *N. sicca* and *N. subflava.*

Neisseria are less nutrition-demanding aerobic bacteria that can grow easily on agar medium. Most *Neisseria* cells

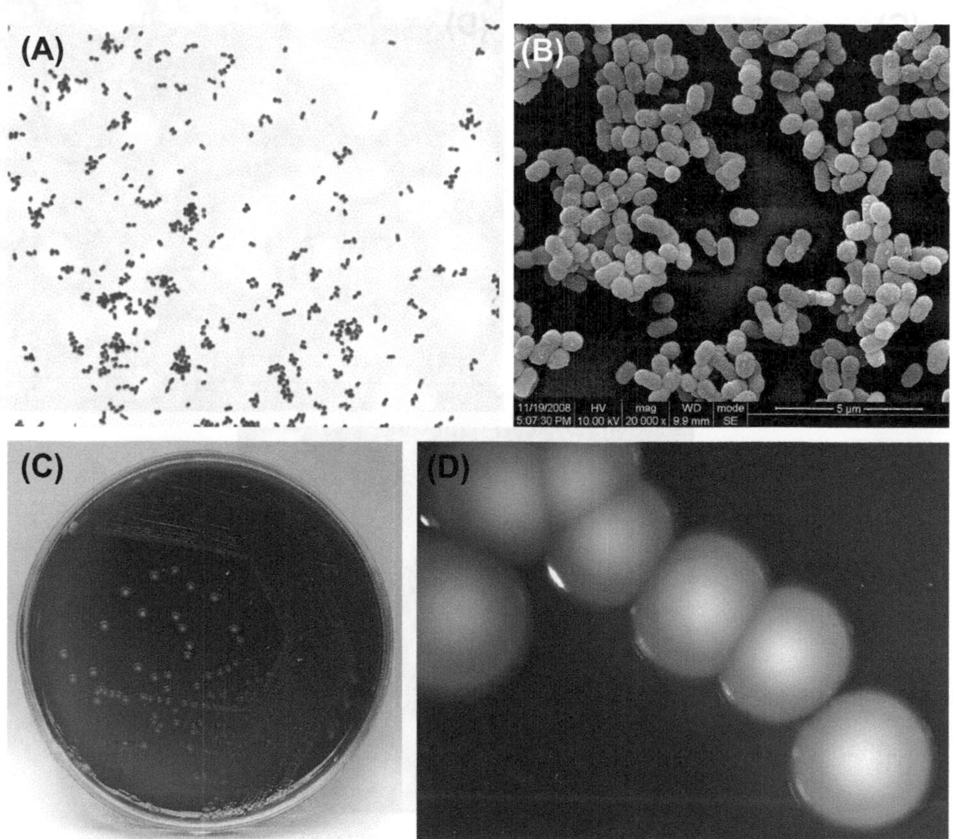

FIGURE 5.8 (A) *Moraxella catarrhalis* cells (Gram stain). (B) *M. catarrhalis* cells (SEM). *M. catarrhalis* cells are spheres or short rods, and are typically kidney-shaped. Nearly kidney-shaped diplococci are observed in clinical specimens. The cells do not produce spores and are nonmotile. Cells stain gram-negative. (C) *M. catarrhalis* colonies (BHI blood agar). (D) *M. catarrhalis* colonies (stereomicroscope).

are spherical, but occasionally short rods are observed with a diameter of 0.6–1.0 μm. Many are arranged in pairs with a flat adjacent surface. *Neisseria* cells may have a capsule and pili, but no endospores and flagella.

5.2.4.1 Neisseria meningitides

Also known as *meningococcus*, this is the pathogen behind meningococcal meningitis. *N. meningitides* are gram-negative cocci (Figure 5.9(A) and (B)). They are aerobic, but grow better under 5–8% CO_2. Growth requires nutrient agar with blood serum or blood. Typical colonies are shown in Figure 5.9(C), (D), and (E).

The genomic GC content is 50–52% (by Tm method). The type strain is ATCC13077.

The colony size of *N. meningitides* depends on the medium on which they are grown. After 18–24 h culture, the colony diameter reaches 1.0–1.5 mm on the surface of nutrient-rich blood agar. They appear as round, smooth, transparent, dew-like colonies without hemolysis. The colonies grow best on chocolate blood agar. On Mueller–Hinton agar, colonies are round, smooth, shiny, and translucent.

5.3 MYCOPLASMA

Mycoplasma can live independently with no cell wall. They are the smallest prokaryotic microbial cells and can be isolated from normal human and animal respiratory mucosa. In the oral cavity, *M. pneumoniae*, *M. oralis*, *M. salivalis*, and *M. nominis* can be isolated.

Mycoplasma are the smallest of all prokaryotic microbial cells, Figure 5.10(A) and (B).

Mycoplasma have high nutritional demands and can grow on PPLO agar with beef heart infusion and 10–20% horse serum. The serum provides *Mycoplasma* with the cholesterol and long-chain fatty acids required for growth. The optimum pH for *Mycoplasma* culture is pH 7.8–8.0. Cells may die when the pH drops below pH 7.0.

Mycoplasma are aerobic or facultative anaerobic microorganisms, but they usually grow better in an aerobic environment. The best culture environment for initial isolation is atmospheric conditions supplemented with 5% CO_2 or anaerobic conditions with 5% CO_2 and 95% N_2.

Mycoplasma colonies are small and can only be observed under a light microscope at low magnification or a dissecting microscope. Characteristic colonies are shown in Figure 5.10(C) and (D).

Mycoplasma and L type bacteria are similar: (1) they both lack a cell wall and the cell is pleomorphic; (2) they can both pass through an antimicrobial filter. The main differences between the two are: (1) *Mycoplasma* are independent microbes and L-type bacteria are variants of normal bacterial cells that have a cell wall (most L-type cells will revert to their original form when the induction factor is eliminated); (2) *Mycoplasma* growth requires cholesterol (10–20% serum in the medium) while the growth

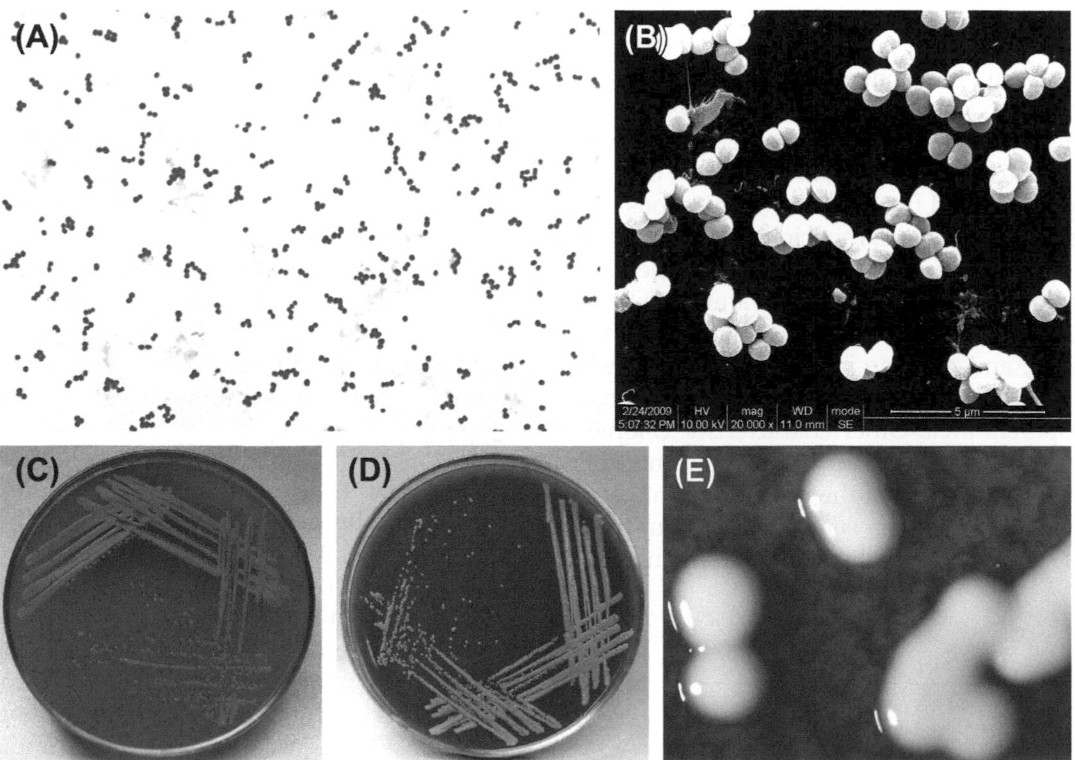

FIGURE 5.9 (A) *Neisseria meningitides* cells (Gram stain). (B) *N. meningitides* cells (SEM). *N. meningitides* cells are ovoid or spherical, kidney-shaped or bean-shaped diplococci, Gram stain is negative. (C) *N. meningitides* colonies (BHI blood agar). (D) *N. meningitides* colonies (BHI chocolate agar). (E) *N. meningitides* colonies (stereomicroscope).

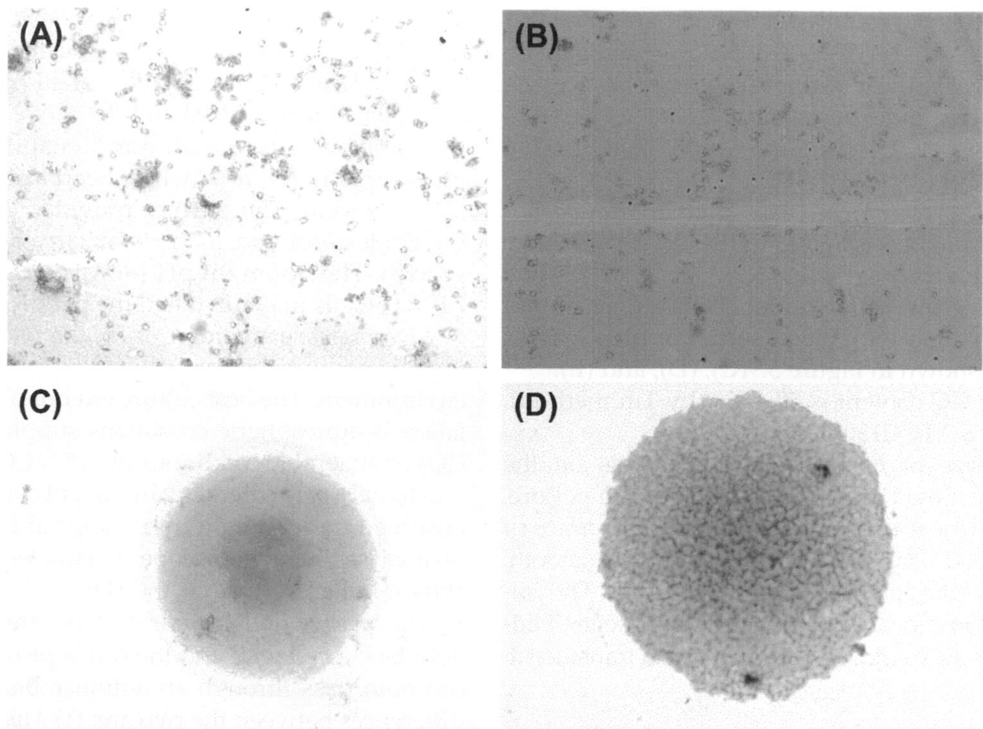

FIGURE 5.10 (A) *Mycoplasma pneumoniae* cells (Gram stain). The signet-ring-shaped cell of *Mycoplasma* is gram-negative, and the size of the cell is 0.2–0.3 μm and is normally smaller than 1.0 μm. Cells have no cell wall. Protein and lipids form the outer cell membrane and the cells are obviously pleomorphic, with spherical, rod-like, bar-like, and filamentous morphologies visible under the microscope. The typical cell is shaped like a signet ring. Cells are gram-negative, light purple with Giemsa stain. (B) *M. pneumoniae* signet ring cell (Giemsa stain). *M. pneumoniae* cell is light purple with Giemsa stain. (C) *M. pneumoniae* omelet-like colonies. (D) Mulberry-shaped colonies of *M. pneumonia*.

of L-type bacteria does not; (3) L-type bacteria fade easily after Diane staining, while *Mycoplasma* do not fade easily.

Mycoplasma usually colonize the throat, bacterial biofilm, or tartar found in the oral cavity. *Mycoplasma pneumoniae* is one of the common causes of acute respiratory infections. Although previous studies reported that *Mycoplasma* can be separated from root canal infections, gingivitis, and periodontitis clinical specimens, the distribution and pathogenicity in the oral cavity are still unclear.

Typical colonies of the *Mycoplasma* are omelet-like in shape. Colony diameter is about 10–15 μm, the center is round, opaque, and extends into the medium, and the edge of the colony is thin and flat, forming a transparent or semitransparent area. Other characteristic colonies of the *Mycoplasma* include mulberry-shaped colonies of *M. pneumoniae*. *Mycoplasma* from the mouth and saliva form visible comet-like colonies in semisolid medium (mostly in the middle and bottom of the culture medium).

5.4 FUNGI

5.4.1 *Saccharomyces*

The family *Saccharomycetaceae* belongs to the phylum *Ascomycota* and includes the common genera *Candida* and *Saccharomyces*. Common species that can be detected in the mouth include *C. albicans*, *S. tropicalis*, *S. candidaglabrata*, *S. parapsilosis*, *S. krusei*, *S. guillermondii*, and *S. dulbilin*, the most commonly detected of these is *C. albicans*.

Saccharomyces is a common symbiotic yeast that inhabits the gastrointestinal tract, respiratory tract, and the vaginal mucosa and can only infect the host under specific conditions. Therefore, they are known as conditional pathogens.

The thallus of this group of yeast is circular or ovoid and consists of a cell wall, cell membrane, cytoplasm, and nucleus. They reproduce by budding. Spores elongate into the germ tube but do not detach from the thallus and form long pseudohyphae.

5.4.1.1 Candida albicans

C. albicans is a common fungus found in the oral cavity and can be detected in the oral cavity of 30–35% of healthy adults. The percentage of detection is even higher in the neonate oral cavity. The main sites of colonization are the mucosa (buccal mucosa, palate, etc.), saliva, oral prosthetics, dentures, etc. *C. albicans* can cause acute or chronic oral candidiasis such as pseudomembranous candidiasis (thrush), denture stomatitis, and *Candida* leukoplakia. Oral candidiasis and other oral *Candida* infections are often secondary infections in AIDS patients.

The *Candida* cell is large and spherical and stains gram-positive (Figure 5.11(A)). The cell in its budding

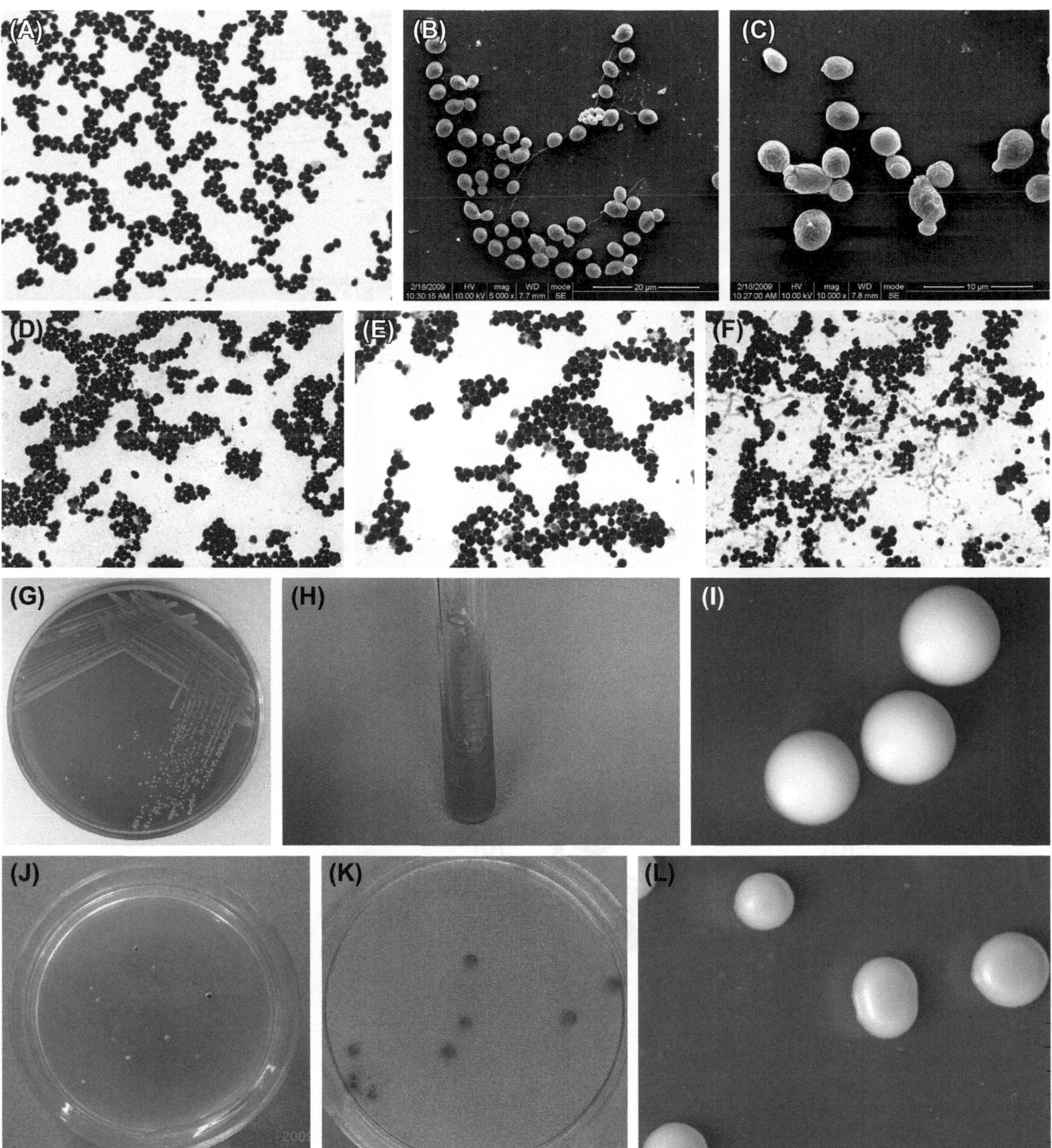

FIGURE 5.11 (A) *Candida albicans* cells (gram-positive). (B) *C. albicans* cells (SEM). (C) *C. albicans* cells (SEM). The yeast cell is 2 × 4 mm in size, ovoid or spherical, gram-positive. The budding cell is visible under SEM. (D) *Candida tropicalis* cells (Gram stain). (E) *C. glabrata* cells (Gram stain). (F) *C. krusei* cells (Gram stain). (G) *C. albicans* colonies (Sabouraud agar). (H) *C. albicans* colonies (Sabouraud agar slant). (I) *C. albicans* colonies (Sabouraud agar, stereomicroscope). *C. albicans* grow well in Sabouraud agar and form milk white, smooth surface, and softer yeast-like colony and yeast smell. When the hatching time is extended, it can form rough surface and faviform or folding rough colony. (J) *C. albicans* colonies (CHROMagar *Candida* agar). (K) *C. albicans* green colonies (CHROMagar *Candida* agar). (L) *C. albicans* colonies (CHROMagar *Candida* agar, stereomicroscope). (M) *Candida tropicalis* blue colonies (CHROMagar *Candida* agar). (N) *C. glabrata* purple colonies (CHROMagar *Candida* agar). (O) *C. glabrata* purple colonies (CHROMagar *Candida* agar, stereomicroscope). (P) *C. krusei* pink colonies (CHROMagar *Candida* agar). (Q) *C. krusei* pink colonies (CHROMagar *Candida* agar, stereomicroscope). (R) *C. albicans* sprouting spores and germ tube (crystal violet stain). (S) *C. albicans* sprouting spores and germ tube (lactophenol stain). Germ tube formation assay involves inoculating the isolated strain on 0.5–1 ml human or sheep serum. Incubate for 2–4 h at 37 °C, stain with crystal violet stain or lactophenol stain, and observe under the microscope. Sprouting spores and germ tubes are visible under oil immersion. (T) *C. albicans* colonies (1% Twain cornmeal agar). (U) Pseudohypha and chlamydospores of *C. albicans* (1% Twain cornmeal agar, crystal violet stain). (V) Pseudohypha and chlamydospores of *C. albicans* (1% Twain cornmeal agar, lactophenol staining).

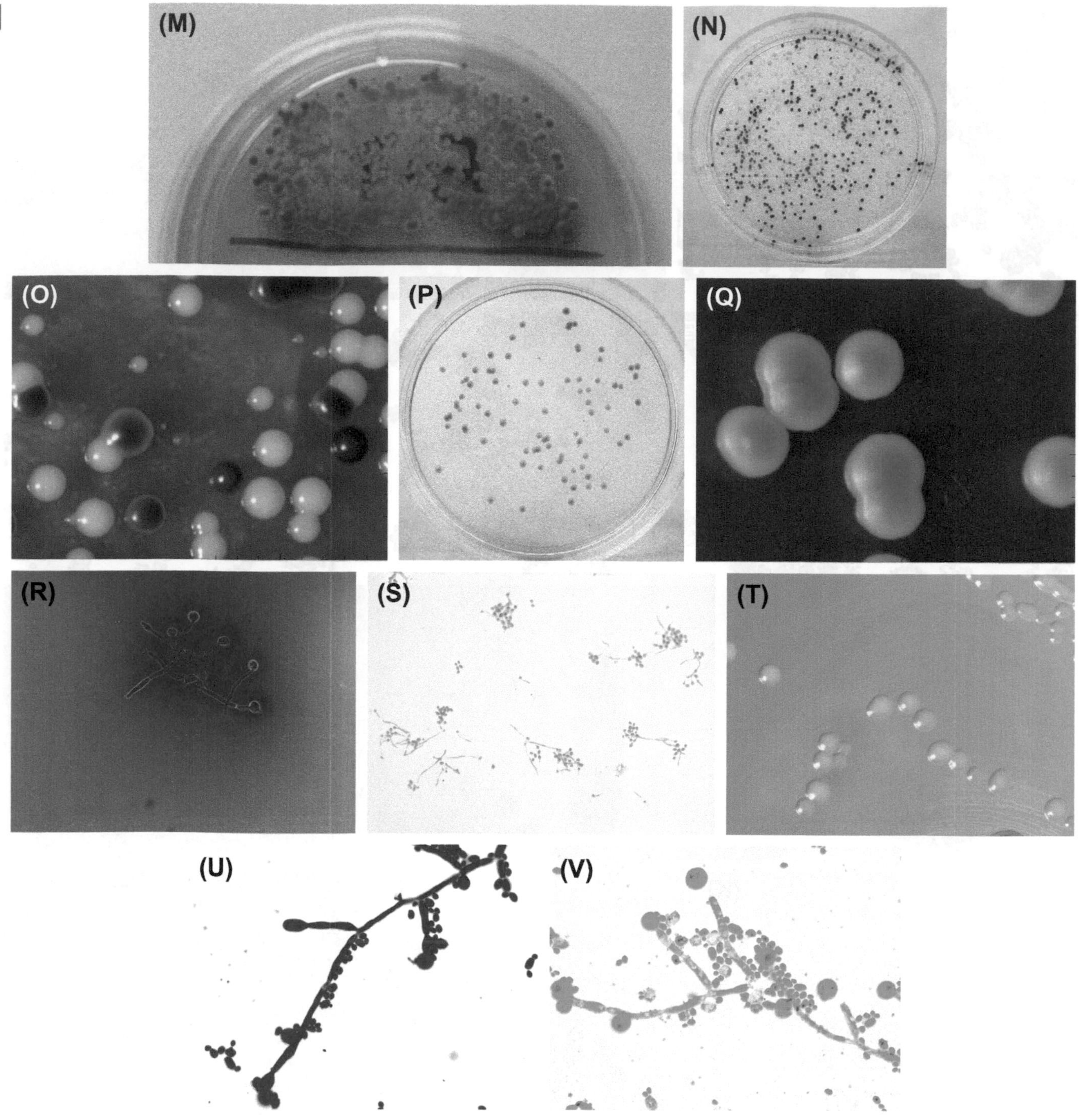

FIGURE 5.11 Cont'd.

form can be visualized by SEM (Figure 5.11(B) and (C)). Other strains are shown in Figure 5.11(D), (E), and (F).

Candida is an aerobic microorganism and grows best in temperatures ranging from 30–37 °C, with 24–48 h incubation time. Sabouraud agar is the standard medium used for its cultivation (Figure 5.11(G), (H), and (I)). CHROMagar *Candida* agar is a commonly used selective medium, as different strains form colonies with different colors on the surface of the medium, which can be used to identify strains from a clinic sample (Figure 5.11(J)). *C. albicans* colonies are shown in Figure 5.11(K) and (L). *Candida tropicalis* colonies and cells are shown in Figure 5.11(M). *Candida glabrata* colonies and cells are shown in Figure 5.11(N) and (O). *Candida krusei* colonies and cells are shown in Figure 5.11(P) and (Q).

Germ tube formation assay (Figure 5.11(R) and (S)) and thick-walled spore formation assay (Figure 5.11(T), (U), and (V)) are commonly used to identify the *C. albicans* strains in a sample. *Candida* can ferment glucose, maltose, and sucrose and produces acid, but does not ferment lactose.

5.5 VIRUS

5.5.1 *Herpes Simplex Virus*

Herpesvirus is an enveloped DNA virus of medium size. More than 100 strains have been discovered. Of these, herpes simplex virus (HSV), varicella-zoster virus, cytomegalovirus, and Epstein–Barr virus are related to human oral mucosal infections. These viruses share an icosahedral protein nucleocapsid formed by 162 subunits that protects the double-stranded linear DNA. There is a lipoprotein envelope surrounding the nucleocapsid. The outer diameter of the virus is approximately 150–200 nm.

HSV is the type virus of the herpesviruses. It causes vesicular lesions of the skin and mucosa and is a common virus found in humans. Research has shown that 30–90% of humans possess antibodies against HSV, indicating that they have previously been infected. HSV can easily invade ectodermal tissues including neurons, skin, and mucosal layers.

The double-stranded DNA contained within the HSC core encodes the genetic information necessary for virus production. The DNA consists of two fragments, one long and one short, linked by a covalent bond. The long fragment makes up 82% of the viral genetic material and the short fragment, 18%. The percentage GC content is approximately 68%, which is fairly high. The nucleocapsid is made up of protein subunits, each of which measures 9.5 × 12.5 nm, with a 4 nm hole at the center. The envelope is a lipid structure rich in proteoglycans and lipoproteins, within which six viral antigens are located: gB, gC, gD, gE, gG, and gH.

In humans, HSV infection is very common and patients and healthy carriers can be sources of infection.

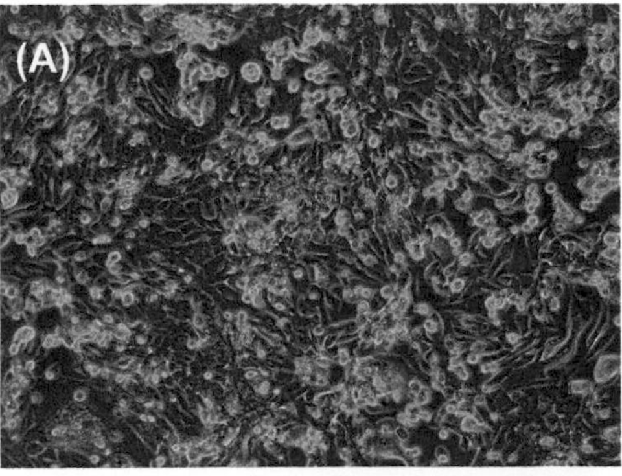

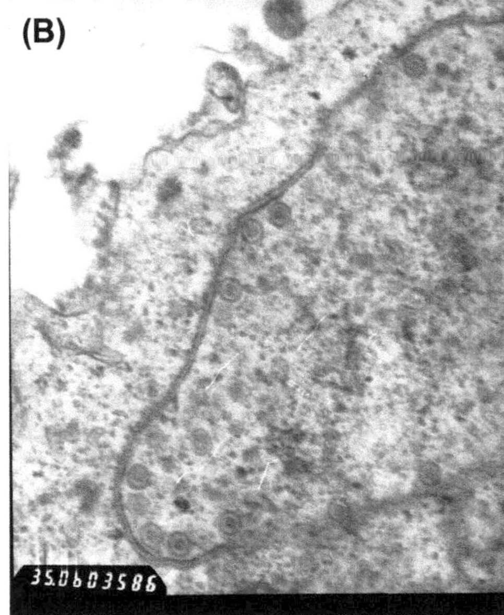

FIGURE 5.12 (A) Proliferation of HSV in Vero cells (inverted microscope). (B) Proliferation of HSV in Vero cells (TEM).

Infection is mainly caused by droplets or coming into contact with saliva or other herpesvirus-containing fluids. The fetus can be infected by passage through the birth canal. In oral leukoplakia, HSV can be isolated from oral squamous cell carcinoma, and there is evidence to support that HSV may be an important factor in oral precancerous lesions or squamous cell carcinoma. The proliferation of HSV in Vero cells is shown in Figure 5.12(A) and (B).

maltose, and sucrose and produces acid but does not ferment lactose.

5.5 VIRUS

5.5.1 Herpes Simplex Virus

Edwards Brothers Malloy
Ann Arbor MI. USA
February 10, 2015